ASE Test Preparation

Automotive Technician Certification Series

Engine Repair (A1)
5th Edition

DELMAR
CENGAGE Learning

Australia • Brazil • Japan • Korea • Mexico • Singapore • Spain • United Kingdom • United States

DELMAR
CENGAGE Learning™

ASE Test Preparation: Automotive Technician Certification Series, Engine Repair (A1), 5ᵗʰ Edition

Vice President, Technology and Trades Professional Business Unit: Gregory L. Clayton

Director, Professional Transportation Industry Training Solutions: Kristen L. Davis

Product Manager: Katie McGuire

Editorial Assistant: Danielle Filippone

Director of Marketing: Beth A. Lutz

Marketing Manager: Jennifer Barbic

Senior Production Director: Wendy Troeger

Production Manager: Sherondra Thedford

Content Project Management: PreMediaGlobal

Senior Art Director: Benj Gleeksman

© 2012, 2006, 2004, 2001 Delmar, Cengage Learning

ALL RIGHTS RESERVED. No part of this work covered by the copyright herein may be reproduced, transmitted, stored, or used in any form or by any means graphic, electronic, or mechanical, including but not limited to photocopying, recording, scanning, digitizing, taping, Web distribution, information networks, or information storage and retrieval systems, except as permitted under Section 107 or 108 of the 1976 United States Copyright Act, without the prior written permission of the publisher.

For product information and technology assistance, contact us at **Cengage Learning Customer & Sales Support, 1-800-354-9706**

For permission to use material from this text or product, submit all requests online at **www.cengage.com/permissions**.
Further permissions questions can be e-mailed to **permissionrequest@cengage.com**

ISBN-13: 978-1-111-12703-9

ISBN-10: 1-111-12703-4

Delmar
Executive Woods
5 Maxwell Drive
Clifton Park, NY 12065
USA

Cengage Learning is a leading provider of customized learning solutions with office locations around the globe, including Singapore, the United Kingdom, Australia, Mexico, Brazil, and Japan. Locate your local office at: **www.cengage.com/global**

Cengage Learning products are represented in Canada by Nelson Education, Ltd.

For more information on transportation titles available from Delmar Cengage Learning, please visit our website at **www.trainingbay.cengage.com**

Visit our corporate website at **www.cengage.com**

Notice to the Reader

Publisher does not warrant or guarantee any of the products described herein or perform any independent analysis in connection with any of the product information contained herein. Publisher does not assume, and expressly disclaims, any obligation to obtain and include information other than that provided to it by the manufacturer. The reader is expressly warned to consider and adopt all safety precautions that might be indicated by the activities described herein and to avoid all potential hazards. By following the instructions contained herein, the reader willingly assumes all risks in connection with such instructions. The publisher makes no representations or warranties of any kind, including but not limited to, the warranties of fitness for particular purpose or merchantability, nor are any such representations implied with respect to the material set forth herein, and the publisher takes no responsibility with respect to such material. The publisher shall not be liable for any special, consequential, or exemplary damages resulting, in whole or part, from the readers' use of, or reliance upon, this material.

Table of Contents

Preface

Delmar, a part of Cengage Learning, is very pleased that you have chosen to use our ASE Test Preparation Guide to help prepare yourself for the Engine Repair (A1) ASE certification examination. This guide is designed to help you prepare for your actual exam by providing you with an overview and an introduction of the testing process, introducing you to the task list for the Engine Repair (A1) certification exam, giving you an understanding of what knowledge and skills you are expected to have in order to successfully perform the duties associated with each task area, and providing you with several preparation exams designed to emulate the live exam content in hopes of accessing your overall exam readiness.

If you have a basic working knowledge of the discipline you are testing for, you will find this book to be an excellent guide, helping you understand the "must know" items needed to successfully pass the ASE certification exam. This manual is not a textbook. Its objective is to prepare the individual who has the existing requisite experience and knowledge to attempt the challenge of the ASE certification process. This guide cannot replace the hands-on experience and theoretical knowledge required by ASE to master the vehicle repair technology associated with this exam. If you are unable to understand more than a few of the preparation questions and their corresponding explanations in this book, it could be that you require either more shop-floor experience or further study.

This book begins by providing an overview of, and introduction to, the testing process. This section outlines what we recommend you do to prepare, what to expect on the actual test day, and overall methodologies for your success. This section is followed by a detailed overview of the ASE task list to include explanations of the knowledge and skills you must possess to successfully answer questions related to each particular task. After the task list, we provide six sample preparation exams for you to use as a means of evaluating areas of understanding, as well as areas requiring improvement in order to successfully pass the ASE exam. Delmar is the first and only test preparation organization to provide so many unique preparation exams. We enhanced our guides to include this support as a means of providing you with the best preparation product available. Section 6 of this guide includes the answer keys for each preparation exam, along with the answer explanations for each question. Each answer explanation also contains a reference back to the related task or tasks that it assesses. This will provide you with a quick and easy method for referring back to the task list whenever needed. The last section of this book contains blank answer sheet forms you can use as you attempt each preparation exam, along with a glossary of terms.

OUR COMMITMENT TO EXCELLENCE

Thank you for choosing Delmar, Cengage Learning for your ASE test preparation needs. All of the writers, editors, and Delmar staff have worked very hard to make this test preparation guide second to none. We feel confident that you will find this guide easy to use and extremely beneficial as you prepare for your actual ASE exam.

Delmar, Cengage Learning has sought out the best subject-matter experts in the country to help with the development of *ASE Test Preparation: Automotive Technician Certification Series, Engine*

Repair (A1), 5th Edition. Preparation questions are authored and then reviewed by a group of certified subject-matter experts to ensure the highest level of quality and validity to our product.

If you have any questions concerning this guide or any guide in this series, please visit us on the web at **http://www.trainingbay.cengage.com**.

For online test preparation solutions for ASE certifications, please visit us on the web at **http://www.techniciantestprep.com** to learn more.

ABOUT THE AUTHOR

Charles Ginther has had ASE Master status in automotive since 1987, with 24 years of experience in the automotive industry. Charles is an automotive educator, an adjunct faculty member for 8 years, and full time faculty member for 16 years in post-secondary education. Service includes: officer for North American Council of Automotive Teachers (NACAT) and founding secretary/ treasurer for North American Council of Automotive Teachers South Carolina Chapter (NACAT/SC).

ABOUT THE SERIES ADVISOR

Mike Swaim has been an Automotive Technology Instructor at North Idaho College, Coeur d'Alene, Idaho, since 1978. He is an Automotive Service Excellence (ASE) Certified Master Technician since 1974 and holds a Lifetime Certification from Mobile Air Conditioning Society. He served as Series Advisor to all nine of the 2011 Automobile/Light Truck Certification Tests (A Series) of Delmar, Cengage ASE Test Preparation titles, and is the author of *ASE Test Preparation: Automobile Certification Series, Undercar Specialist Designation (X1), 5th Edition.*

The History and Purpose of ASE

ASE began as the National Institute for Automotive Service Excellence (NIASE). It was founded as a non-profit, independent entity in 1972 by a group of industry leaders with the single goal of providing a means for consumers to distinguish between incompetent and competent technicians. It accomplishes this goal through the testing and certification of repair and service professionals. Though it is still known as the National Institute for Automotive Service Excellence, it is now called "ASE" for short.

Today, ASE offers more than 40 certification exams in automotive, medium/heavy duty truck, collision repair and refinish, school bus, transit bus, parts specialist, automobile service consultant, and other industry-related areas. At this time, there are more than 385,000 professionals nationwide with current ASE certifications. These professionals are employed by new car and truck dealerships, independent repair facilities, fleets, service stations, franchised service facilities, and more.

ASE's certification exams are industry-driven and cover practically every on-highway vehicle service segment. The exams are designed to stress the knowledge of job-related skills. Certification consists of passing at least one exam and documenting two years of relevant work experience. To maintain certification, those with ASE credentials must be re-tested every five years.

While ASE certifications are a targeted means of acknowledging the skills and abilities of an individual technician, ASE also has a program designed to provide recognition for highly qualified repair, support, and parts businesses. The Blue Seal of Excellence Recognition Program, allows businesses to showcase their technicians and their commitment to excellence. One of the requirements of becoming Blue Seal recognized is that the facility must have a minimum of 75 percent of their technicians ASE certified. Additional criteria apply, and program details can be found on the ASE website.

ASE recognized that educational programs serving the service and repair industry also needed a way to be recognized as having the faculty, facilities, and equipment to provide a quality education to students wanting to become service professionals. Through the combined efforts of ASE, industry, and education leaders, the non-profit organization entitled the National Automotive Technicians Education Foundation (NATEF) was created in 1983 to evaluate and recognize academic programs. Today more than 2,000 educational programs are NATEF certified.

For additional information about ASE, NATEF, or any of their programs, the following contact information can be used:

> National Institute for Automotive Service Excellence (ASE)
>
> 101 Blue Seal Drive S.E.
>
> Suite 101
>
> Leesburg, VA 20175
>
> Telephone: 703-669-6600
>
> Fax: 703-669-6123
>
> Website: **www.ase.com**

Participating in the National Institute for Automotive Service Excellence (ASE) voluntary certification program provides you with the opportunity to demonstrate you are a qualified and skilled professional technician who has the "know-how" required to successfully work on today's modern vehicles.

EXAM ADMINISTRATION

> *Note:* After November 2011, ASE will no longer offer paper and pencil certification exams. There will be no Winter testing window in 2012, and ASE will offer and support CBT testing exclusively starting in April 2012.

ASE provides computer-based testing (CBT) exams, which are administered at test centers across the nation. It is recommended that you go to the ASE website at http://www.ase.com and review the conditions and requirements for this type of exam. There is also an exam demonstration page that allows you to personally experience how this type of exam operates before you register.

CBT exams are available four times annually, for two-month windows, with a month of no testing in between each testing window:

- January/February—Winter testing window
- April/May—Spring testing window
- July/August—Summer testing window
- October/November—Fall testing window

Please note, testing windows and timing may change. It is recommended you go to the ASE website at *http://www.ase.com* and review the latest testing schedules.

UNDERSTANDING TEST QUESTION BASICS

ASE exam questions are written by service industry experts. Each question on an exam is created during an ASE-hosted "item-writing" workshop. During these workshops, expert service representatives from manufacturers (domestic and import), aftermarket parts and equipment manufacturers, working technicians, and technical educators gather to share ideas and convert them into actual exam questions. Each exam question written by these experts must then survive review by all members of the group. The questions are designed to address the practical application of repair and diagnosis knowledge and skills practiced by technicians in their day-to-day work.

After the item-writing workshop, all questions are pre-tested and quality-checked on a national sample of technicians. Those questions that meet ASE standards of quality and accuracy are included in the scored sections of the exams; the "rejects" are sent back to the drawing board or discarded altogether.

Depending on the topic of the certification exam, you will be asked between 40 and 80 multiple-choice questions. You can determine the approximate number of questions you can expect to be asked during the Engine Repair (A1) certification exam by reviewing the task list in Section 4 of this book. The five-year recertification exam will cover this same content; however, the number of questions for each content area of the recertification exam will be reduced by approximately one-half.

> *Note:* Exams may contain questions that are included for statistical research purposes only. Your answers to these questions will not affect your score, but since you do not know which ones they are, you should answer all questions in the exam.

Using multiple criteria, including cross-sections by age, race, and other background information, ASE is able to guarantee that exam questions do not include bias for or against any particular group. A question that shows bias toward any particular group is discarded.

TEST-TAKING STRATEGIES

Before beginning your exam, quickly look over the exam to determine the total number of questions that you will need to answer. Having this knowledge will help you manage your time throughout the exam to ensure you have enough available to answer all of the questions presented. Read through each question completely before marking your answer. Answer the questions in the order they appear on the exam. Leave the questions blank that you are not sure of and move on to the next question. You can return to those unanswered questions after you have finished the others. These questions may actually be easier to answer at a later time once your mind has had additional time to consider them on a subconscious level. In addition, you might find information in other questions that will help you recall the answers to some of them.

Multiple-choice exams are sometimes challenging because there are often several choices that may seem possible, or partially correct, and therefore it may be difficult to decide on the most appropriate answer choice. The best strategy, in this case, is to first determine the correct answer before looking at the answer options. If you see the answer you decided on, you should still be careful to examine the other answer options to make sure that none seem more correct than yours. If you do not know or are not sure of the answer, read each option very carefully and try to eliminate those options that you know are incorrect. That way, you can often arrive at the correct choice through a process of elimination.

If you have gone through the entire exam, and you still do not know the answer to some of the questions, *then guess*. Yes, guess. You then have at least a 25 percent chance of being correct. While your score is based on the number of questions answered correctly, any question left blank, or unanswered, is automatically scored as incorrect.

There is a lot of "folk" wisdom on the subject of test taking that you may hear about as you prepare for your ASE exam. For example, there are those who would advise you to avoid response options that use certain words such as *all, none, always, never, must,* and *only,* to name a few. This, they claim, is because nothing in life is exclusive. They would advise you to choose response options that use words that allow for some exception, such as *sometimes, frequently, rarely, often, usually, seldom,* and *normally.* They would also advise you to avoid the first and last option (A or D) because exam writers, they feel, are more comfortable if they put the correct answer in the middle (B or C) of the choices. Another recommendation often offered is to select the option that is either shorter or longer than the other three choices because it is more likely to be correct. Some would advise you to never change an answer since your first intuition is usually correct. Another area of "folk" wisdom focuses specifically on any repetitive patterns created by your question responses (e.g., A, B, C, A, B, C, A, B, C).

Many individuals may say that there are actual grains of truth in this "folk" wisdom, and whereas with some exams, this may prove true, it is not relevant in regard to the ASE certification exams. ASE validates all exam questions and test forms through a national sample of technicians, and only those questions and test forms that meet ASE standards of quality and accuracy are included in the scored sections of the exams. Any biased questions or patterns are discarded altogether, and therefore, it is highly unlikely you will experience any of this "folk" wisdom on an actual ASE exam.

PREPARING FOR THE EXAM

Delmar, Cengage Learning wants to make sure we are providing you with the most thorough preparation guide possible. To demonstrate this, we have included hundreds of preparation questions in this guide. These questions are designed to provide as many opportunities as possible to prepare you to successfully pass your ASE exam. The preparation approach we recommend and outline in this book is designed to help you build confidence in demonstrating what task area content you already know well while also outlining what areas you should review in more detail prior to the actual exam.

We recommend that your first step in the preparation process should be to thoroughly review Section 3 of this book. This section contains a description and explanation of the type of questions you'll find on an ASE exam.

Once you understand how the questions will be presented, we then recommend that you thoroughly review Section 4 of this book. This section contains information that will help you establish an understanding of what the exam will be evaluating, and specifically, how many questions to expect in each specific task area.

As your third preparatory step, we recommend you complete your first preparation exam, located in Section 5 of this book. Answer one question at a time. After you answer each question, review the answer and question explanation information located in Section 6. This section will provide you with instant response feedback, allowing you to gauge your progress, one question at a time, throughout this first preparation exam. If after reading the question explanation you do not feel you understand the reasoning for the correct answer, go back and review the task list overview (Section 4) for the task that is related to that question. Included with each question explanation is a clear identifier of the task area that is being assessed (e.g., Task A.1). If at that point you still do not feel you have a solid understanding of the material, identify a good source of information on the topic, such as an educational course, textbook, or other related source of topical learning, and do some additional studying.

After you have completed your first preparation exam and have reviewed your answers, you are ready to complete your next preparation exam. A total of six practice exams are available in Section 5 of this book. For your second preparation exam, we recommend that you answer the questions as if you were taking the actual exam. Do not use any reference material or allow any interruptions in order to get a feel for how you will do on the actual exam. Once you have answered all of the questions, grade your results using the Answer Key in Section 6. For every question that you gave an incorrect answer to, study the explanations to the answers and/or the overview of the related task areas. Try to determine the root cause for missing the question. The easiest thing to correct is learning the correct technical content. The hardest things to correct are behaviors that lead you to an incorrect conclusion. If you knew the information but still got the question incorrect, there is likely a test-taking behavior that will need to be corrected. An example of this would be reading too quickly and skipping over words that affect your reasoning. If you can identify what you did that caused you to answer the question incorrectly, you can eliminate that cause and improve your score.

Here are some basic guidelines to follow while preparing for the exam:

- Focus your studies on those areas you are weak in.
- Be honest with yourself when determining if you understand something.
- Study often but for short periods of time.
- Remove yourself from all distractions when studying.
- Keep in mind that the goal of studying is not just to pass the exam; the real goal is to learn.
- Prepare physically by getting a good night's rest before the exam, and eat meals that provide energy but do not cause discomfort.
- Arrive early to the exam site to avoid long waits as test candidates check in.
- Use all of the time available for your exams. If you finish early, spend the remaining time reviewing your answers.
- Do not leave any questions unanswered. If absolutely necessary, guess. All unanswered questions are automatically scored as incorrect.

Here are some items you will need to bring with you to the exam site:

- A valid government or school-issued photo ID
- Your test center admissions ticket
- A watch (not all test sites have clocks)

> *Note:* Books, calculators, and other reference materials are not allowed in the exam room. The exceptions to this list are English-Foreign dictionaries or glossaries. All items will be inspected before and after testing.

WHAT TO EXPECT DURING THE EXAM

When taking a CBT exam, as soon as you are seated in the testing center, you will be given a brief tutorial to acquaint you with the computer-delivered test prior to taking your certification exam(s). The CBT exams allow you to select only one answer per question. You can also change your answers as many times as you like. When you select a second answer choice, the CBT will automatically unselect your first answer choice. If you want to skip a question to return to later, you can utilize the "flag" feature, which will allow you to quickly identify and review questions whenever you are ready. Prior to completing your exam, you will also be provided with an opportunity to review your answers and address any unanswered questions.

TESTING TIME

Each individual ASE CBT exam has a fixed time limit. Individual exam time will vary based upon exam area, and will range anywhere from a half hour to two hours. You will also be given an additional 30 minutes beyond what is allotted to complete your exams to ensure you have adequate time to perform all necessary check-in procedures, complete a brief CBT tutorial, and potentially complete a post-test survey.

You can register for and take multiple CBT exams during one testing appointment. The maximum time allotment for a CBT appointment is four and a half hours. If you happen to register for so many exams that you will require more time than this, your exams will be scheduled into multiple appointments. This could mean that you have testing on both the morning and afternoon of the same day, or they could be scheduled on different days, depending on your personal preference and the test center's schedule.

It is important to understand that if you arrive late for your CBT test appointment, you will not be able to make up any missed time. You will only have the scheduled amount of time remaining in your appointment to complete your exam(s).

Also, while most people finish their CBT exams within the time allowed, others might feel rushed or not be able to finish the test, due to the implied stress of a specific, individual time limit allotment. Before you register for the CBT exams, you should review the number of exam questions that will be asked along with the amount of time allotted for that exam to determine whether you feel comfortable with the designated time limitation or not.

As an overall time management recommendation, you should monitor your progress and set a time limit you will follow with regard to how much time you will spend on each individual exam question. This should be based on the total number of questions you will be answering.

Also, it is very important to note that if for any reason you wish to leave the testing room during an exam, you must first ask permission. If you happen to finish your exam(s) early and wish to leave the testing site before your designated session appointment is completed, you are permitted to do so only during specified dismissal periods.

UNDERSTANDING HOW YOUR EXAM IS SCORED

You can gain a better perspective about the ASE certification exams if you understand how they are scored. ASE exams are scored by an independent organization having no vested interest in ASE or in the automotive industry. With CBT exams, you will receive your exam scores immediately.

Each question carries the same weight as any other question. For example, if there are 50 questions, each is worth 2 percent of the total score.

Your exam results can tell you:

- Where your knowledge equals or exceeds that needed for competent performance, or
- Where you might need more preparation.

Your ASE exam score report is divided into content "task" areas; it will show the number of questions in each content area and how many of your answers were correct. These numbers provide information about your performance in each area of the exam. However, because there may be a different number of questions in each content area of the exam, a high percentage of correct answers in an area with few questions may not offset a low percentage in an area with many questions.

It should be noted that one does not "fail" an ASE exam. The technician who does not pass is simply told "More Preparation Needed." Though large differences in percentages may indicate problem areas, it is important to consider how many questions were asked in each area. Since each exam evaluates all phases of the work involved in a service specialty, you should be prepared in each area. A low score in one area could keep you from passing an entire exam. If you do not pass the exam, you may take it again at any time it is scheduled to be administered.

There is no such thing as average. You cannot determine your overall exam score by adding the percentages given for each task area and dividing by the number of areas. It doesn't work that way because there generally are not the same number of questions in each task area. A task area with 20 questions, for example, counts more toward your total score than a task area with 10 questions.

Your exam report should give you a good picture of your results and a better understanding of your strengths and areas needing improvement for each task area.

Understanding not only what content areas will be assessed during your exam, but how you can expect exam questions to be presented will enable you to gain the confidence you need to successfully pass an ASE certification exam. The following examples will help you recognize the types of question styles used in ASE exams and assist you in avoiding common errors when answering them.

Most initial certification tests are made up of between 40 and 80 multiple-choice questions. The five-year recertification exams will cover the same content as the initial exam; however, the actual number of questions for each content area will be reduced by approximately one-half. Refer to Section 4 of this book for specific details regarding the number of questions to expect during the initial Engine Repair (A1) certification exam.

Multiple-choice questions are an efficient way to test knowledge. To correctly answer them, you must consider each answer choice as a possibility, and then choose the answer choice that *best* addresses the question. To do this, read each word of the question carefully. Do not assume you know what the question is asking until you have finished reading the entire question.

About 10 percent of the questions on an actual ASE exam will reference an illustration. These drawings contain the information needed to correctly answer the question. The illustration should be studied carefully before attempting to answer the question. When the illustration is showing a system in detail, look over the system and try to figure out how the system works before you look at the question and the possible answers. This approach will ensure that you do not answer the question based upon false assumptions or partial data, but instead have reviewed the entire scenario being presented.

MULTIPLE-CHOICE/DIRECT QUESTIONS

The most common type of question used on an ASE exam is the direct multiple-choice style question. This type of question contains an introductory statement, called a stem, followed by four options: three incorrect answers, called distracters, and one correct answer, the key.

When the questions are written, the point is to make the distracters plausible to draw an inexperienced technician to inadvertently select one of them. This type of question gives a clear indication of the technician's knowledge.

Here is an example of a direct style question:

1. Which of the following would be used to measure crankshaft main bearing clearance on diesel engine?

   A. Dial indicator
   B. Plastigauge®
   C. Outside micrometer
   D. Dial caliper

TASK B.15

Answer A is incorrect. A dial indicator is used to measure end-play or run out. It would not be useful to measure crankshaft main bearing clearance.

Answer B is correct. Plastigauge is the most common method of measuring crankshaft main bearing clearance.

Answer C is incorrect. An outside micrometer is used to measure the outside dimension of a part; however, alone it cannot measure crankshaft main bearing clearance.

Answer D is incorrect. A dial caliper is used to measure outside depth or inside dimensions, but would not be an effective tool to measure crankshaft main bearing clearance.

COMPLETION QUESTIONS

A completion question is similar to the direct question except the statement may be completed by any one of the four options to form a complete sentence.

Here is an example of a completion question:

TASK E.2

2. A diesel engine has an accelerator pedal position diagnostic trouble code. The technician wiggles the wiring harness while observing the accelerator pedal position sensor voltage with the scan tool. The sensor voltage changes. The most likely cause of the diagnostic trouble code is a:

   A. Faulty scan tool.
   B. Faulty ECM.
   C. Faulty accelerator pedal position sensor wiring.
   D. Faulty ECM power supply.

Answer A is incorrect. If the voltage value changed while moving the wiring harness, there is no reason to believe the scan tool is faulty.

Answer B is incorrect. If the voltage value changed while moving the wiring harness, there is no reason to believe the ECM is faulty.

Answer C is correct. If the voltage value changed while moving the wiring harness, the most likely cause is the wiring harness.

Answer D is incorrect. If all other items associated with the ECM are normal and the voltage value changed while moving the wiring harness, the most likely cause is the wiring harness.

TECHNICIAN A, TECHNICIAN B QUESTIONS

This type of question is usually associated with an ASE exam. It is, in fact, two true-false statements grouped together, such as: "Technician A says…" and "Technician B says…", followed by "Who is correct?"

In this type of question, you must determine whether either, both, or neither of the statements are correct. To answer this type of question correctly, you must carefully read each technician's statement and judge it on its own merit.

Sometimes this type of question begins with a statement about some analysis or repair procedure. This statement provides the setup or background information required to understand the conditions about which Technician A and Technician B are talking, followed by two statements about the cause of the concern, proper inspection, identification, or repair choices.

Analyzing this type of question is a little easier than the other types because there are only two ideas to consider, although there are still four choices for an answer.

Again, Technician A, Technician B questions are really double true-or-false questions. The best way to analyze this type of question is to consider each technician's statement separately. Ask yourself, "Is A true or false? Is B true or false?" Once you have completed this individual evaluation of each answer choice, you will have successfully determined the correct answer choice for the question, "Who is correct?".

An important point to remember is that an ASE Technician A, Technician B question will never have Technician A and B directly disagreeing with each other. That is why you must evaluate each statement independently.

An example of a Technician A/Technician B style question looks like this:

TASK A.4

1. A noise is coming from the accessory drive on the front of a diesel engine. Technician A says the serpentine belt can be removed to help determine if it is the source of the noise. Technician B says water dripped on the belt can help determine if the belt is the source of the noise. Who is correct?

    A. A only

    B. B only

    C. Both A and B

    D. Neither A nor B

Answer A is incorrect. Technician B is also correct.

Answer B is incorrect. Technician A is also correct.

Answer C is correct. Both Technicians are correct. The belt can be removed and the engine run for a short time to determine if the belt is the source of the noise. Also, if the noise disappears when a few drops of water are put on the running belt, the technician knows the belt is the source of the noise.

Answer D is incorrect. Both Technicians are correct.

EXCEPT QUESTIONS

Another type of question type used on the ASE exams contains answer choices that are all correct except for one. To help easily identify this type of question, whenever it is presented in an exam, the word "EXCEPT" will always be displayed in capital letters. Furthermore, a cautionary statement will alert you to the fact that the next question is different from the ones otherwise found in the exam. With the EXCEPT type of question, only one *incorrect* choice will actually be listed among the options, and that incorrect choice will be the key to the question. That is, the incorrect statement is counted as the correct answer for that question.

Be careful to read these question types slowly and thoroughly; otherwise, you may overlook what the question is actually asking and answer the question by selecting the first correct statement.

An example of this type of question would appear as follows:

TASK A.3

1. A diesel engine is being checked for an engine oil leak. All of the following could be used to help locate the source of the leak EXCEPT:

    A. Black light.

    B. White powder.

    C. Vacuum gauge.

    D. Oil dye.

Answer A is incorrect. A black light can be used to help locate the source of a leak.

Answer B is incorrect. White powder can be used to help locate the source of a leak.

Answer C is correct. A vacuum gauge is not used to locate engine oil leaks.

Answer D is incorrect. Oil dye can be used to help locate an engine oil leak.

LEAST LIKELY QUESTIONS

LEAST LIKELY questions are similar to EXCEPT questions. Look for the answer choice that would be the LEAST LIKELY cause (most incorrect) of the described situation. To help easily identify these types of question, whenever they are presented in an exam the words "LEAST LIKELY" will always be displayed in capital letters. In addition, you will be alerted before a LEAST LIKELY question is posed. Read the entire question carefully before choosing your answer.

An example of this type of question is shown here:

TASK D.12

1. A vehicle equipped with a diesel engine overheats when pulling a trailer. Which of the following would be the LEAST LIKELY cause?

    A. Slipping fan clutch
    B. Seized fan clutch
    C. Restricted charge air cooler
    D. Restricted radiator

Answer A is incorrect. A slipping fan clutch may not fully engage and would fail to provide sufficient air flow across the radiator to keep the engine cool.

Answer B is correct. A seized fan clutch would run all the time; this may cause a low power complaint, but would not cause the engine to overheat.

Answer C is incorrect. A restricted charge air cooler would also restrict the air flow across the radiator. This could result in an engine overheating condition.

Answer D is incorrect. A restricted radiator could result in an overheated engine.

SUMMARY

The question styles outlined in this section are the only ones you will encounter on any ASE certification exam. ASE does not use any other types of question styles, such as fill-in-the-blank, true/false, word-matching, or essay. ASE also will not require you to draw diagrams or sketches to support any of your answer selections, although any of the described question styles may include illustrations, charts, or schematics to clarify a question. If a formula or chart is required to answer a question, it will be provided for you.

Task List Overview

INTRODUCTION

This section of the book outlines the content areas or *task list* for this specific certification exam, along with a written overview of the content covered in the exam.

The task list describes the actual knowledge and skills necessary for a technician to successfully perform the work associated with each skill area. This task list is the fundamental guideline you should use to understand what areas you can to expect to be tested on, as well as how each individual area is weighted to include the approximate number of questions you can expect to be given for that area during the ASE certification exam. It is important to note that the number of exam questions for a particular area is to be used as a guideline only. ASE advises that the questions on the exam may not equal the number listed on the task list. The task lists are specifically designed to tell you what ASE expects you to know how to do and to help prepare you to be tested.

Similar to the role this task list will play in regard to the actual ASE exam, Delmar, Cengage Learning has developed six preparation exams, located in Section 5 of this book, using this task list as a guide. It is important to note that although both ASE and Delmar, Cengage Learning use the same task list as a guideline for creating these test questions, none of the test questions you will see in this book will be found in the actual, live ASE exams. This is true for any test preparatory material you use. Real exam questions are *only* visible during the actual ASE exams.

Task List at a Glance

The Engine Repair (A1) task list focuses on six core areas, and you can expect to be asked a total of approximately 50 questions on your certification exam, broken out as outlined here:

 A. General Engine Diagnosis (15 questions)
 B. Cylinder Head and Valve Train Diagnosis and Repair (10 questions)
 C. Engine Block Diagnosis and Repair (10 questions)
 D. Lubrication and Cooling Systems Diagnosis and Repair (8 questions)
 E. Fuel, Electrical, Ignition, and Exhaust Systems Inspection and Service (7 questions)

Based upon this information, the graph shown here is a general guideline demonstrating which areas will have the most focus on the actual certification exam. This data may help you prioritize your time when preparing for the exam.

> *Note:* The actual number of questions you will be given on the ASE certification exam may vary slightly from the information provided in the task list, as exams may contain questions that are included for statistical research purposes only. Don't forget that your answers to these research questions will not affect your score.

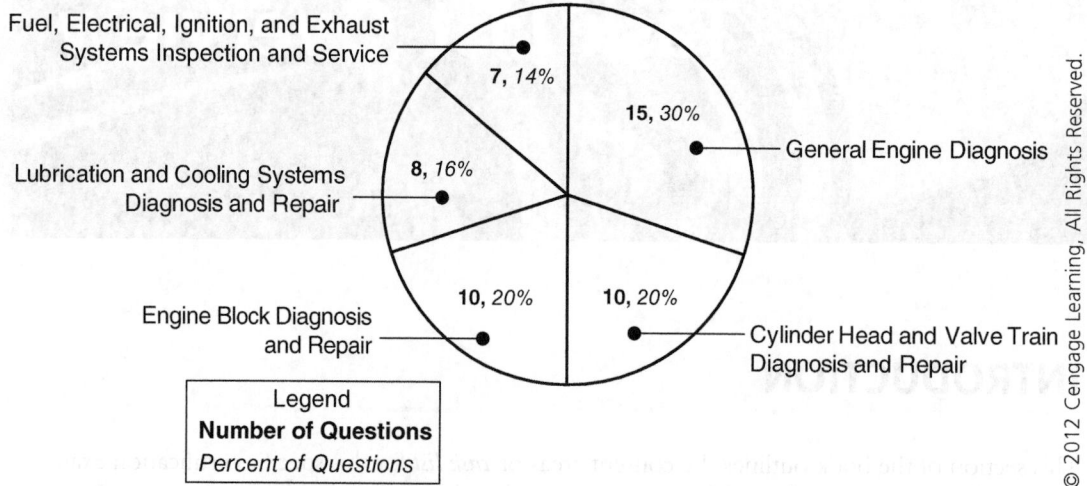

Fuel, Electrical, Ignition, and Exhaust Systems Inspection and Service — *7, 14%*

General Engine Diagnosis — *15, 30%*

Lubrication and Cooling Systems Diagnosis and Repair — *8, 16%*

Engine Block Diagnosis and Repair — *10, 20%*

Cylinder Head and Valve Train Diagnosis and Repair — *10, 20%*

Legend
Number of Questions
Percent of Questions

© 2012 Cengage Learning, All Rights Reserved.

ENGINE REPAIR (TEST A1)

A. General Engine Diagnosis (15 Questions)

1. Verify driver's complaint and/or road test vehicle; determine necessary action.

In the modern automotive repair shop environment, a Customer Service Specialist, or Service Consultant, is usually responsible for collecting information from the customer to assist the technician in resolving a customer's concern. Armed with this information, his skills, and a fair amount of intuition, the technician must determine the cause of the particular concern.

One of the most important tools a technician can utilize is the pre-test drive. During this drive, he may be able to determine the problem and other conditions that may contribute to or relate the concern. After verifying the customer's concern, the technician must determine the most efficient diagnostic path to take. Most experienced technicians will check with available information systems to see if the problem they are diagnosing is described in a technical service bulletin or service campaign. In some instances, it may be necessary to have the customer drive the vehicle with the technician as a passenger so the customer can create the exact scenario for the complaint.

Once the problem is found and resolved, a final test drive allows the technician to confirm that the problem has indeed been resolved and that no other problems have come to light after the repair.

2. Determine if no-crank, cranks but will not start, or hard starting condition is an engine mechanical problem, or is caused by another vehicle subsystem.

If the starter fails to crank the engine, the problem may range from a faulty starter motor to broken components inside the engine. If no sounds come from the starter motor when it is activated, first disable the ignition system, and then attempt to rotate the crankshaft pulley by hand in the normal direction of rotation. If the crankshaft can be rotated freely through two complete revolutions, a diagnosis of the vehicles starting system is the next step.

If you are unable to rotate the crankshaft by hand, the engine may be hydrostatically locked or have broken internal components. To check for hydrostatic lock, remove all the spark plugs and attempt to rotate the crankshaft again. If oil or coolant squirts from the spark plug holes, this indicates a bad head gasket, warped cylinder head or block, a cracked cylinder head or block, or the customer ran through deep water. Water squirting from the spark plug holes could be an indication the customer went through a deep puddle and drew in water through the air intake.

If the crankshaft cannot be rotated at all with the spark plugs removed, or cannot be rotated through at least one complete revolution, the engine may be seized or have broken internal parts. Pull the dipstick and check the crankcase oil level. If oil does not register on the dipstick, it is possible that the pistons are seized in their bores or the connecting rods are seized to the crankshaft. If the oil level is sufficient, a broken component may have lodged between moving parts inside the cylinder block, thus preventing the parts from rotating.

Many overhead camshaft (OHC) engines are non-freewheeling or "interference" engines. On these engines, a no-crank condition may be caused by piston-to-valve contact. This is a common occurrence when a timing belt slips or breaks, but it may also occur on engines fitted with a timing chain and sprockets. On many belt-driven OHC engines, it is possible to easily loosen or remove part of the timing belt cover. Do this, if possible, and check for obvious signs of belt failure.

If the customer states that the starter cranks the engine but it will not start (or takes a long time to start), confirm that the valve train is operating properly before attempting to crank the engine yourself. If the timing belt or chain is broken or jumped, additional cranking may cause severe engine damage.

A no-start or hard starting complaint can be caused by a faulty ignition, malfunctioning fuel/emission control systems, or severely worn internal engine components. These complaints can also be caused by broken or slipped valve train timing components, especially on free-wheeling engines. A broken timing device may cause some cylinders to have good compression while others have none. A slipped timing device may result in all cylinders having low compression. To determine if the belt or chain is functioning properly, rotate the crankshaft by hand while observing the distributor rotor or camshaft. If these components fail to rotate with the crankshaft, the timing belt or chain is broken. It is common for timing belts that have not been replaced at original equipment manufacturer (OEM) intervals to have their teeth torn off by the crank sprocket. The timing belt will appear normal during the visual inspection, but fail to rotate as the engine is turned over. If they do rotate with the crankshaft, confirm proper indexing of the rotor or camshaft to determine if the belt or chain has slipped. Rotate the crankshaft until the piston in cylinder #1 is at top dead center (TDC) on the compression stroke. Then check distributor rotor or camshaft position to make sure that it is correct.

In the real world, most technicians will check for the presence of spark before checking for fuel or mechanical issues. The process for checking spark requires using a tool called a "spark tester" in place of one of the spark plugs and cranking the engine to visually check for spark. Some distributorless ignition systems are not compatible with this type of test and may be tested using appropriate secondary ignition lab scope procedures.

Fuel injected engines usually receive high-pressure fuel from an electric pump. To verify that the fuel pump is operating and fuel is reaching the engine, locate the fuel line that supplies fuel to the throttle body or fuel injector rail. Turn the key on/engine off after installing a fuel pressure gauge. Verify that the gauge registers adequate pressure. If the gauge does not register any pressure or registers very low pressure, proceed with the fuel system diagnosis. If fuel pressure is adequate, begin the diagnosis of the fuel injection control system.

3. Inspect engine assembly for fuel, oil, coolant, and other leaks; determine necessary action.

The source of fluid leaks can be difficult to locate. Determining what type of fluid is leaking will reduce the number of possible leak locations.

Engine oil usually leaks from faulty gaskets and seals, but it can also leak from cracked castings, faulty pressure switches or sending units, and loose tapered oil gallery plugs. Oil can leak from an area high on the engine (like a V-type engine intake manifold rear seal) and run down the engine, appearing at the rear of the oil pan. Do not assume that the "wet" area is the source of the leak. Clean the area and run the engine to check for fresh fluid. Also, do not immediately assume that a leaking seal or gasket is faulty. Excessive blow-by or a faulty pressure control valve (PCV) system can pressurize the crankcase, forcing oil past a seal or gasket that is in good condition. Hard-to-find oil leaks can be located by pouring a small quantity of fluorescent dye into the crankcase and running the engine. When an ultraviolet light is shined onto the engine, oil containing the dye will glow to reveal the leak point. White aerosol foot powder may also be used to help find leaks. First, thoroughly degrease the area around the leak. Spray the area with a white aerosol foot powder. Start and operate vehicle then observe the area. The white foot powder provides a bright background to make leak detection a little easier.

Fuel may leak from loose connections or damaged components. Check for loose hose clamps and fuel line fittings. Check hoses for swelling, cracks, and damage from abrasion. Check metal lines for cracks and corrosion. Check for leaking o-ring connections and a leaking fuel pressure regulator (often mounted on the throttle body unit or injector rail).

The soft metal plugs that seal the cooling system channels in the cylinder block and sometimes the cylinder heads are known by many names. We will use core plug in this text but be aware that they are also known as expansion or freeze plugs and that these terms are interchangeable.

Corroded core plugs are common coolant leakage points, as are faulty hoses and water pumps. Also check coolant temperature sensors, sending units, and thermal vacuum switches. Some engines have core plugs at the back of the cylinder block and/or head. If the engine consumes coolant, but you cannot find evidence of leaking coolant, check the engine oil level and condition. Coolant may be leaking into the crankcase. Coolant can also leak into the combustion chambers.

On vehicles equipped with power steering, check the fluid level. Leaking power steering fluid may be mistaken for engine oil or transmission fluid. When checking for oil leaks, note the color of the oil. A brown to black color usually indicates engine oil, red usually indicates automatic transmission fluid, and clear usually indicates power steering fluid.

4. Isolate engine noises and vibrations; determine necessary action.

Different types of engine part failures often make distinctive sounds. First, be sure that the noise is actually coming from the engine. A faulty water pump, alternator, power steering pump, A/C compressor, or air injection pump can make noises that appear to be coming from inside the engine. Loose or broken accessory mounting brackets can also cause noises that sound like engine internal problems. Listen to each of the accessories using a stethoscope to determine if it is the source of a noise. If in doubt, temporarily remove the drive belt from an accessory to prevent it from operating.

A faulty crankshaft main or rod bearing usually makes a knocking sound that is very deep in pitch. Main bearing knock is usually a thumping noise most noticeable when the engine is first started. Connecting rod bearings also cause a heavy knocking sound, and engine oil pressure may also be low, especially at idle. When the cylinder with faulty connecting rod bearings is disabled during a cylinder balance test, the knocking sound will diminish. Loose flywheel bolts may cause a thumping noise at idle. Camshaft bearings usually do not cause a noise unless severely worn.

Worn pistons and cylinders cause a rapping noise while accelerating. When performing a cylinder balance test, piston noise can increase when the faulty cylinder is disabled (the opposite reaction of a bad connecting rod bearing). A piston pin with excessive clearance often makes a "double click" noise when the engine is idling.

Lifters also make a distinctive noise, a loud ticking sound. One way to isolate lifter (or other valve train) noise from connecting rod noise is to remember that the camshaft operates at half of crankshaft speed. It is common for a lifter with excessive leak-down to tick for a few seconds after the engine starts. The noise goes away once full oil pressure is developed.

Vibrations can be caused by engine accessory components, but more often are caused by internal engine failures such as bearings or pistons. Some vibrations will result from a faulty vibration damper. Vibrations caused by faulty vibration dampers can be identified as they typically will be present only at certain RPMs, i.e. 750, 1,500, and 2,250. Vibrations can also be caused by out of time balance shafts. These shafts are chain, belt, or gear driven.

5. Diagnose the cause of excessive oil consumption, coolant consumption, unusual engine exhaust color, and odor; determine necessary action.

Excessive oil consumption can be due to oil leaking from the engine or oil being drawn into the cylinders and burned. Before blaming internal components, be absolutely sure that oil is not leaking from the engine. In some cases, oil leaks only when the engine is running. If necessary, raise the vehicle on a lift while the engine is running to check for leaks. The radiator and/or coolant recovery jug should also be checked for signs of oil leakage into the cooling system.

Oil can enter the cylinders several different ways, including: worn rings; scored cylinder walls; worn valve guides, seals, and stems; worn turbocharger seals; and plugged oil drain passages. As a general rule, an engine that is "burning oil" will emit blue-gray exhaust. This may be more noticeable on acceleration and deceleration. Do not confuse the blue-gray smoke due to oil consumption with the black exhaust that occurs when the air/fuel ratio is too rich.

Plugged oil drain passages in the cylinder head or block can cause excessive oil consumption even when rings and guides are in good condition. To check for this, remove the oil filler cap or another component fitted to a valve cover and start the engine. If the oil level inside the cover rises steadily as the engine runs and reaches the top of the valve guides, the drain passages are clogged. While these passages can usually be cleared of sludge, the sludge is an indication that the engine was poorly maintained. Clearing the passages will probably reduce oil consumption, but the engine may experience other problems in the near future. Oil found in the cooling system can be leaking from an internal oil cooler (in the radiator tank) or from a head gasket on an OHC engine.

Perform compression tests (Task A.8) and cylinder leakage tests (Task A.9) to confirm whether piston rings/cylinders or valve guides are worn.

If the engine is turbocharged, first perform oil consumption diagnosis as though the engine was not turbocharged. While turbos are commonly blamed for excessive oil consumption problems, about half of the turbos returned under warranty are not defective. If oil is found in the turbo compressor housing or intake manifold, check the oil drain from the turbo housing to the block. If it is obstructed, oil under pressure will be forced past the turbo seals and into the engine through the intake tract. Check the PCV system, too. If the PCV valve does not close during "boost" conditions, the crankcase will be pressurized. This may pressurize the turbo oil drain passage, forcing oil into the turbo housing.

Like oil consumption, coolant consumption may be caused by coolant leaking from the cooling system or coolant leaking into the engine (or passenger compartment). First, eliminate external leaks as the cause of coolant consumption by performing cooling system pressure tests (Task D.3). If cooling system pressure drops during the tests but no leaks are found, check the engine oil level and condition. Leaks into the crankcase will raise the oil level. If coolant is being drawn into the combustion chambers, the exhaust will be gray or white. The engine will continue to emit this smoke long after the time it usually takes for moisture to be purged from the exhaust system. On vehicles equipped with electronic engine controls, coolant passing through the exhaust system will "poison" the oxygen sensor.

Some engine problems can be diagnosed by listening to the exhaust pulses at the tailpipe. If all cylinders are firing properly, the exhaust should consist of steady pulses. A puffing noise that occurs at regular intervals usually indicates a cylinder misfire caused by a compression, ignition, or fuel system defect. Puffing noises that occur erratically are usually caused by ignition or fuel system defects. Engine idle speed may also be unsteady.

A high-pitched squealing noise during hard acceleration may be caused by a small leak in the exhaust system, particularly in the exhaust manifolds or exhaust pipe. The leak may also be noticeable at idle as a ticking noise.

Another common engine noise is a high-pitched whistle at idle and low engine speeds. Check for vacuum leaks at the intake manifold gaskets. Also check for cracked or disconnected vacuum hoses. A vacuum leak whistle gradually decreases when the engine is accelerated and the intake vacuum decreases.

A strong sulfur or rotten egg smell coming from the exhaust system of a car fitted with a catalytic converter may indicate a rich air/fuel ratio.

6. Perform engine vacuum tests; determine necessary action.

A vacuum test can be used to help pinpoint the cause of an engine problem. The vacuum gauge should be connected directly to the intake manifold.

On an engine that is performing correctly, the vacuum gauge reading should be between 17 and 22 in. Hg (45 and 28 kPa absolute) and steady with the engine idling. Some abnormal vacuum gauge readings and typical problems associated with them are:

1. A slightly low but steady reading may indicate retarded ignition timing.
2. A very low but steady reading may indicate that the intake manifold has a significant leak.
3. Burned or leaking valves may cause the vacuum gauge to fluctuate.
4. Weak valve springs may result in a vacuum gauge fluctuation.
5. A leaking head gasket may cause a vacuum gauge fluctuation.

6. If the valves are sticking, the vacuum gauge fluctuates.

7. If, when the engine is accelerated and held steady at a higher speed, the vacuum gauge pointer gradually falls, the catalytic converter or other exhaust system components are restricted.

7. Perform cylinder power balance tests; determine necessary action.

In a sound engine, each cylinder contributes equal amounts of power. A common method of isolating a power loss that may be occurring in one or more of the cylinders is the power balance test. In most engines, this test is performed by disabling the spark on a single cylinder for a short time to measure the RPM drop that occurs when that cylinder is not contributing. The ignition should be disabled for the shortest amount of time possible to prevent catalytic converter damage. Cylinders that are not contributing or are weak will show little or no RPM drop when they are disabled. On port fuel injected engines many manufacturers include the ability to perform cylinder balance tests with the power train control module (PCM) and a scanner. Most on-board diagnostics (OBD) II vehicles will quickly pick up on and identify an inefficient cylinder, displaying a code directing to the perceived problem and the affected cylinder.

When a particular cylinder or cylinders fail a power balance test, it becomes necessary to determine if the cause is mechanical: valves, internal engine efficiency, or engine-management related. Engine management issues could include port fuel injectors that leak or are electrically damaged (delivering no fuel at all), faulty ignition components like spark plugs, distributorless ignition systems (DIS) coils, plug wires, etc. The power balance test is usually the precursor to a cylinder compression and/or leak down test. These tests will confirm a mechanical issue and electrical diagnosis using digital storage oscilloscopes or scanners can determine engine management issues.

8. Perform cylinder cranking compression tests; determine necessary action.

The ignition and fuel injection system must be disabled before proceeding with the compression test. During the compression test, the throttle is blocked open and the engine is cranked through four compression strokes for each cylinder. The compression readings are recorded for each stroke and compared to the manufacturer's specifications. Slightly low compression readings in all cylinders are not cause for concern if engine performance is acceptable. Compression readings that vary more than 20 percent (from the highest to the lowest) are cause for concern. A wet compression test should be performed to help pinpoint causes of low compression. To perform a wet test, add several squirts of oil to the cylinder allowing 30 to 60 seconds for the oil to flow around the rings and then retest the compression. If the compression shows great improvement, the most likely causes are piston ring and/or cylinder bore wear. A wet test usually will not seal leaking gaskets or valves and will show little or no improvement in the compression reading if these components are at fault. Compression readings may be interpreted as follows:

■ When the compression readings on all the cylinders are about equal, but significantly lower than specifications, the piston rings or cylinder walls may be worn. If compression in all cylinders is low and the engine spins freely during cranking, check the valve timing. The timing belt or sprocket may have jumped.

■ Low compression readings on one or more cylinders indicate worn rings, leaking valves, a blown head gasket, flat camshaft, or a cracked cylinder head. Performing a leak down test will narrow down the cause of the problem.

■ Low compression readings in two adjacent cylinders are probably due to a leaking head gasket or cracked cylinder head.

■ Zero compression in a cylinder is usually caused by a hole in a piston or a severely burned exhaust valve. If the zero compression reading is caused by a hole in the piston, the engine will have excessive blow-by.

■ Higher than specified compression usually indicates carbon deposits in the combustion chamber.

9. Perform cylinder leakage tests; determine necessary action.

During a cylinder leakage test, a regulated amount of air from the shop air supply is forced into the cylinder while both the exhaust and intake valves are closed. The gauge on the leakage tester indicates the percentage of leakage in the cylinder. A gauge reading of 0 percent indicates that there is no cylinder leakage. If the reading is 100 percent, the cylinder is not holding any air.

If cylinder leakage exceeds 20 percent, determine the source of the leakage by listening for air escaping through the tailpipe, crankcase (via the oil filler cap or PCV valve), and/or the intake tract (throttle body or carburetor). Air escaping from the tailpipe indicates an exhaust valve leak. When the air is coming out of the PCV valve or valve cover opening, the piston rings are leaking. An intake valve is leaking if air is escaping from the top of the throttle body or carburetor. Remove the radiator cap and check the coolant for bubbles, which indicates a leaking head gasket or cracked head. Remove all of the spark plugs. Air escaping from an adjacent spark plug hole could indicate a bad head gasket, or crack in the block or cylinder head.

B. Cylinder Head and Valve Train Diagnosis and Repair (10 Questions)

1. Remove cylinder heads, disassemble, clean, and prepare for inspection.

Remove a cylinder head only when the engine is cold. Removing a warm cylinder head may cause the head to warp, especially if it is made of aluminum.

Remove the cylinder head bolts, loosening the bolts in a sequence opposite that of the tightening sequence. Note and record the positions of special bolts. Remove the cylinder head from the engine. Cylinder heads can be quite heavy, so ask an assistant to help you, especially if the engine is still mounted in the vehicle.

Use a spring compressor to compress the valve springs and then remove the valve locks, or "keepers." Release the compressor and remove the retainer, rotator, spring, and spring seats from the head. Keep all parts in an organizer so they can be returned to their original cylinder. Check the valve stem tips for mushrooming. If it is present, the tip must be dressed with a file before the valves are removed from the head. Remove the valves from the cylinder head and place them in an organizer.

When removing the cylinder head from an OHC design engine, the timing belt or chain must first be disconnected from the camshaft. The procedure for doing this varies from manufacturer to manufacturer. On some engines with a chain-driven camshaft, the camshaft sprocket is unbolted from the cam. The cylinder head assembly is then removed, leaving the chain and sprockets in position on the engine. On some engines with a belt-driven camshaft, the belt tensioner is loosened and the belt is slipped off the camshaft sprocket. On other engines, the timing cover and belt must be completely removed from the engine. If the timing belt is being removed from the engine and will be reused, mark the direction of rotation on the belt. Reinstall the belt so it rotates in the same direction. Never crank the engine after a timing device has been loosened or removed until the cylinder head has also been removed. Cranking an engine while the timing belt is loose or disconnected can cause immediate and serious engine damage.

The basic cylinder head removal procedure varies from manufacturer to manufacturer. On some OHC engines the cylinder head, camshaft(s), and rocker arms (if used) are removed as an assembly after loosening and removing the cylinder head bolts. On some engines the rocker shaft and arm assembly (including the upper half of the cam bearings) must first be removed to access cylinder head mounting bolts. Refer to the appropriate service manual for information.

Cleaning the components for reassembly changed dramatically with the advent of aluminum heads and blocks. Many of the surfaces of these components are polished to near mirror-like qualities. After cleaning these components in the appropriate degreasing solution, carbon buildup may be removed by hand with a soft wire wheel or in an abrasive cabinet like a glass blaster. In either case, care must be taken to insure that gasket surfaces remain in their original condition. Use of refinishing wheels on polished surfaces can cause new gaskets to fail prematurely. The head must be carefully washed after removing deposits and old gasket material to insure that none of the debris restricts passages. Many manufacturers are recommending plastic scrapers and gasket softening chemicals to remove old gaskets. Be sure to find out what the manufacturer recommends to guarantee a quality repair.

2. Visually inspect cylinder heads for cracks, warpage, corrosion, leakage, and the condition of passages; determine needed repairs.

While the cylinder heads from any engine should be carefully inspected, the heads from engines with serious mechanical problems (blown head gasket, coolant consumption, overheating, oil sludging, etc.) should receive special attention. First look at the old head gaskets to determine if a problem area is visible. If one or more is, match the area to the contact area on the cylinder head.

Check the cylinder head for cracks, paying special attention to the combustion chambers and the areas between intake and exhaust valves. An electromagnetic-type tester and iron filings (MAGNAFLUX®) may be used to check for cracks in cast iron heads. A dye penetrant may be used to locate cracks in aluminum heads. Machine shops can usually locate hard-to-find cracks by pressure testing. In this type of test, all coolant passages are blocked using metal plates. The coolant jacket is then filled with compressed air and the head is submerged in a tank of water. Bubbles escaping from the head reveal leak areas.

Use a straightedge and a feeler gauge to check the cylinder head for warpage at several locations. Check the manufacturer's service manual for exact specifications. A cylinder head that is excessively warped must be resurfaced or replaced.

On OHC engines, be sure to check warpage on the cam side of the head before resurfacing and reinstalling the head. If warpage exceeds manufacturer's specifications, the camshaft will bind, flex, and may break.

Inspect the coolant passages in the cylinder head as thoroughly as possible. Shine a flashlight into the passages looking for corrosion, rust, and trapped debris. A cylinder head that shows evidence of severe pitting in the cooling jacket should be replaced.

3. Inspect and repair damaged threads where allowed; install core and gallery plugs.

When removing or installing a component, carefully inspect all bolts, studs, holes, and nuts for damage to threads. Stripped, cross-threaded, nicked, rolled, and rusty threads can produce errors in the torque values. Use compressed air to remove foreign matter from holes. Obstructions lodged in holes, such as metal and liquids, can cause a bolt to bottom out and reach torque value without obtaining the required clamping force. Examine the shank of bolts and studs for signs of twisting or over-torque fracturing. Replace defective bolts and studs. Ensure that nuts are not distorted or cracked, and replace self-locking and defective nuts. Clean and dress threads in holes. If threads are damaged beyond use, drill and tap the hole and install an approved HeliCoil thread insert. Use an approved stud extractor to remove broken studs and bolts. Core plugs are stamped metal plugs; these plugs tend to deteriorate from the inside out, therefore condition cannot be determined by an external observation. These plugs are always replaced during cylinder head overhaul. They will also be replaced while the engine is in service if they are leaking. Core plugs are typically metal cups driven into place with the correct driver tool. The technician will normally coat the outer lip of the plug with sealer prior to installing. There are aftermarket rubber expansion plugs available that can be used when the plug is being replaced with the engine still in the vehicle. Oil gallery plugs can be either a tapered pipe plug or straight thread o-ring type. The tapered pipe plug should have a light coat of sealant applied prior to installation; the straight thread o-ring type should not. Both types of plugs should be installed using a torque wrench.

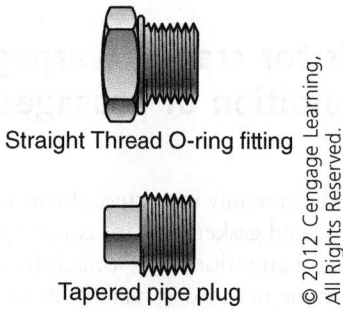

Straight Thread O-ring fitting

Tapered pipe plug

© 2012 Cengage Learning, All Rights Reserved.

4. Inspect, test, and verify valve springs for squareness, pressure, and free height comparison; replace as necessary.

A valve spring tester is used to measure valve spring pressure or tension. Valve springs must be carefully checked to see that the free height of the spring meets specifications and that the spring does not lean to one side. The spring must stand exactly perpendicular when it sits on a flat surface. The spring must be able to attain a specified pressure when at rest. This is called the valve spring installed height. It must also be able to attain specific pressure at maximum valve lift. A spring tester is used to measure these pressures at their respective heights. In most cases, if a spring fails it will not be able to develop adequate pressure. This can cause the valves to "float" or bounce when the engine is at higher revs. Low spring pressure often leads to a spring breaking and the valve falling into the cylinder. In the more unusual event that the spring generates too much pressure, it can cause excessive wear on the cam, followers, or lifters.

5. Inspect valve spring retainers, rotators, locks/keepers, and lock grooves.

Valve spring retainers and locks/keepers must be checked for wear, scoring, or damage. When any of these conditions are present, replace the components.

Valve rotators can be inspected before cylinder head removal and disassembly since they usually cannot be taken apart. Rotators can be located on top of the valve spring (built into the spring retainer) or between the valve spring and the cylinder head. To test a rotator, mark the top of the spring retainer with a dab of paint. Then start the engine and run it at about 1,500 rpm. The retainer should slowly rotate. The direction of rotation is not important. If the retainer does not rotate, replace the rotator.

In some cases, a defective rotator can be diagnosed with the cylinder head removed. Check the valve stem tip wear pattern. A shallow groove or channel running across the tip indicates that the valve has not been rotating. Replace the rotator on any valve with this condition.

Inspect the valve lock/keeper grooves machined into the valve stems. Look for damage and wear, particularly for round shoulders. If the shoulders are uneven or rounded, replace the valve. Observe the depth of the valve lock/keeper in the valve spring retainer. If they are uneven, it may indicate wear that could cause them to pull through the retainer. Valve lock/keeper failures can cause severe engine damage.

6. Replace valve stem seals.

It is not necessary to remove the cylinder head to service many valve train components. This includes the valve springs, oil seals, retainers, and valve locks. A claw-type valve spring compressor can be used to compress the valve spring while the head is mounted on the block, but will require that the shop air supply is connected to the cylinder through the spark plug hold. This air pressure will hold the valves in place when the valve springs are removed, preventing the valves from dropping into the cylinder.

On engines equipped with umbrella-type valve seals, slip the new seals over the valve stems. Work carefully to avoid damaging new seals on valve stem lock grooves. A damaged seal will cause excessive oil consumption. Some seals come with an installation tool. The tool is simply a short plastic sleeve that is slipped over the tip of each valve stem before the seal is installed. The sleeve extends down far enough to cover the lock grooves. After installing each seal, push it down against the top of the valve guide and remove the installation tool.

On engines equipped with positive-type valve seals, the installation procedure is the same with one important difference. Positive-type seals must be pushed down over the top of the valve guide. Each seal has some sort of retaining device to keep it attached to the guide. Some seals have a flat, circular spring that wraps around the seal; some use garter springs to hold the seal in place; others have a molded-in ridge on their inner diameter that mates with a groove machined onto the valve guide. Whatever the retaining method, be sure that the valve guide seal is securely attached to the guide. Some manufacturers specify that a special tool should be used to drive the positive seal onto the valve guide. For further information, refer to a service manual for the vehicle.

7. Inspect valve guides for wear; check valve stem-to-guide clearance; determine needed repairs.

Valve guides should be measured near the top, center, and bottom using a hole gauge. Measure the valve stem diameter with a micrometer in the same three positions, and subtract the stem readings from the guide measurements to obtain the clearance. An alternate method for measuring stem-to-guide clearance is to install the valve in the guide with the valve about $\frac{1}{8}$ inch (3.18 mm) off its seat. Mount a dial indicator against the valve margin or against the valve stem below the lock grooves. Move the valve from side to side while observing the clearance reading on the dial indicator. Divide the reading by two to obtain stem-to-guide clearance.

If clearance exceeds specifications, the valve guides may be replaced or bored out, and a thin-wall liner installed. Excessive valve stem-to-guide clearance may result in an improper valve seating and lower compression. Increased oil consumption may result from excessive valve stem-to-guide clearance.

8. Inspect valves and valve seats; determine needed repairs.

Although most shops no longer perform valve and seat reconditioning, technicians still need to be able to inspect the condition of valves and seats to determine the cause of failure. Older engines and many large displacement light truck engines are more susceptible to wear. The valves in these engines are larger than small engines so they tend to have more trouble holding their shape and dissipating heat into the valve seats. In these larger valves it is not uncommon to see the exhaust valves tulip or crack under hard use. Here again, aluminum heads eliminate most of these conditions due to their ability to draw excessive heat out of the exhaust valves. The downside is that the conditions that caused the valve damage will now, more than likely, damage the cylinder head instead. Engines with multiple valves and smaller displacement engines demonstrate excellent valve and seat life. In most cases, the valves and seats will last the life of the engine. Common failures are valve damage when timing belts break in interference engines and carbon buildup on the valve head that causes poor contact with the seat and loss of compression.

When inspecting valves and seats we are looking for obvious changes in the shape of the valve. The seat-to-valve contact area is critical to good sealing. The valve train is designed to lift the valve and gently set it down. If excessive valve lash occurs in the valve train the result can be accelerated valve seat wear. Any irregularity in this surface can cause a valve to not seal, resulting in cylinder misfire. The valves must seal completely for the cylinder to reach full pressure during both the compression and power strokes.

9. Check valve spring installed (assembled) height and valve stem height; determine needed repairs.

Measure the installed valve stem height from the spring seat surface on the cylinder head to the valve stem tip. If the stem installed height is greater than specifications, the valve stem is stretched or too much material has been removed from the valve face or seat. Install a new valve and measure stem height again. If the measurement is still excessive, replace the seat or cylinder head. Excessive valve stem height moves the plunger downward in a hydraulic valve lifter and may cause valve train components to bottom out.

Measure the installed valve spring height from the lower edge of the top retainer to the spring seat. If this measurement is excessive, install shims between the bottom of the valve spring and the top of the spring seat surface on the cylinder head. Excessive installed valve spring height reduces valve spring tension, which may result in valve float and cylinder misfiring at higher speeds.

▦ 10. Inspect pushrods, rocker arms, rocker arm pivots, and shafts for wear, bending, cracks, looseness, and blocked oil passages; repair or replace as required.

Pushrods should be inspected for a bent condition and wear on the ends. Roll the pushrod on a level surface to check for a bent condition. Bent pushrods usually indicate interference in the valve train, such as a sticking valve, improper valve adjustment, or mechanical interference due to improper valve timing. If the pushrod has an oil passage to provide oil to the rocker arm, make sure that the passage is not obstructed.

Worn rocker arms, shafts, or pivots cause improper valve adjustment and make a clicking noise in the valve train. Check rocker arm shafts for wear and scoring in the rocker arm contact area. Check the shafts for cracks, bending, and loose/leaking oil passage plugs (if fitted). Check rocker arms for scoring at the pivot area and valve stem tip contact area. Worn rocker arms should be replaced.

▦ 11. Inspect and replace hydraulic or mechanical lifters/lash adjusters.

Technicians usually diagnose hydraulic valve train problems from the outside in. In most cases, the customer is complaining of a ticking or clicking noise in the engine. We use our stethoscope to pinpoint the location of the noise and perform the necessary disassembly to arrive at the problem component. It is important to understand how the hydraulic lash adjuster, this includes the traditional V8-type lifter, functions. These components are built to amazingly small tolerances often .0001 of an inch. They have a spring in them that holds the valve train components in place. This spring is not responsible for controlling valve lash, however; this is a function of engine oil pressure. The hydraulic lash adjuster maintains a preload on valve train components at all times when the engine is running.

Hydraulic lash adjusters range in size from about one inch in diameter in cam-in-block engines to as small as .25 of an inch in many of the smaller multi-valve engines. Due to their very close tolerances, they are not very compatible with dirty oil, which is the main cause for failure. In applications where the lifter (lash adjuster) rides directly on the camshaft, there are two designs the flat tappet and the roller design.

The flat tappet design starts life with a slightly convex bottom that encourages it to spin as the angled face of the camshaft raises and lowers it. This motion keeps oil moving over the cam and lifter contact surfaces. When inspecting this type for damage, look for dull surfaces, pitting, or concave contact surfaces. If this kind of damage is present, the camshaft will have to be replaced as well.

The roller lifter was originally used in racing engines because of its ability to follow large and complex camshaft lobe profiles. Due to the fact that a cylinder rolls on the camshaft surface, the contact area between the cam and lifter is very small and has significantly less friction than the flat tappet design. OEMs began using this design to reduce internal friction and net better gas mileage. The hydraulic portion of this design is the same as all the others. The component that rolls will generally not fail unless it fails to roll. Here again watch for signs of pitting or for components that do not roll easily. The roller dictates that the lifter cannot spin so the cam and roller are made of steel. In most applications a used roller lifter that has not failed can be installed on a new camshaft. This is not possible with a flat tappet design where new lifters may be installed on an old cam, but used lifters cannot be installed on a new cam.

When a hydraulic failure occurs in any of the three designs, the unit is replaced with a new component. In some rare instances, the lash adjuster can fail in such a way that it holds the valve open. This can cause valve damage, particularly on exhaust valves.

12. Adjust valves on engines with mechanical or hydraulic lifters.

Valve lash adjusting procedures and mechanisms vary from manufacturer to manufacturer. On some engines valves are adjusted while the engine is cold. On others, the engine should be at operating temperature. Refer to the appropriate service manual for instructions.

On some engines with mechanical valve lifters, the rocker arms have an adjustment screw and a locknut on the valve stem end of the arm. Other engines use an interference fit screw without a locknut. The adjustment procedure usually involves rotating the crankshaft to position the piston in the cylinder being adjusted at top dead center (TDC) on the compression stroke. Feeler gauges are then inserted between the adjusting screw and the valve stem. If clearance is excessive, the adjustment screw locknut (if equipped) is loosened and the adjustment screw is turned. When clearance is correct, a feeler gauge of the correct thickness will slide between the adjusting screw and the valve stem with a light push fit. Tighten the locknut (if equipped) when clearance is correct.

Some OHC engines fitted with mechanical valve lifters have removable metal pads in each lifter or spring retainer. Clearance is measured by placing feeler gauges between the cam lobe and the lifter or spring retainer while the piston is at TDC on the compression stroke. Pads are available in various thicknesses to adjust the clearance to specifications. Another variation of this method will not have removable pads and the entire valve lifter must be replaced with one that has the correct thickness.

Some valve trains have hydraulic valve lifters and individual rocker arm pivots retained with self-locking nuts. These valve trains require an initial adjustment of the rocker arm nut to position the lifter plunger. With the valve closed, loosen the rocker arm nut until there is clearance between the end of the rocker arm and the valve stem. Slowly turn the rocker arm nut clockwise until a zero lash condition exists. This occurs when the lifter is not preloaded but all of the components have just come in contact with each other. The next step is to apply the preload specified by the manufacturer. In most applications, this is the equivalent of .030–.060 of an inch of lifter preload. It is important to follow the manufacturer's specs here as the lifter must be able to compensate for expansion and contraction of valve train components during cold to fully warm engine operation.

The valve train on engines with hydraulic lifters and stud-mounted rocker arms can also be adjusted while the engine is running. After removing the valve covers, install oil shrouds on the rocker arms to prevent oil from splashing onto the exhaust manifolds and other nearby parts. Start the engine and loosen the rocker retaining nut until a clicking noise begins. Then slowly tighten the nut just until the clicking noise stops. From this point, slowly tighten the nut about $1/4$ turn at a time for the specified number of turns. Wait a few seconds between each $1/4$ turn to allow the lifter to leak down. Turning the adjusting screw too much at one time or too far can cause piston-to-valve contact.

13. Inspect and replace camshaft(s) (includes checking drive gear wear and backlash, end-play, sprocket and chain wear, overhead cam drive sprocket(s), drive belt(s), belt tension, tensioners, camshaft reluctor ring/tone-wheel, and variable valve timing components).

When the camshaft gear teeth mesh directly with the crankshaft gear teeth, gear backlash may be measured with a dial indicator positioned against one of the camshaft gear teeth. Rock the cam gear back and forth and note the maximum reading on the indicator.

Some engines equipped with a timing chain and sprockets are fitted with a hydraulic tensioner. The tensioner uses pressurized oil from the lubrication system to eliminate timing chain play. Some manufacturers recommend measuring the installed length of the tensioner to determine chain wear. If the tensioner length exceeds the manufacturer's specifications, replace the timing chain.

On many V-type camshafts in block engines, a timing mark on the crankshaft sprocket must be aligned with a timing mark on the camshaft sprocket before the camshaft sprocket and chain are installed. Timing chain stretch and wear may be measured on these engines with a socket and flex handle installed on one of the camshaft sprocket retaining bolts. Rock the camshaft sprocket back and forth without moving the crankshaft gear and measure the movement on one of the chain link pins on the camshaft sprocket.

Engines fitted with a timing belt should have the belt replaced at the mileage intervals recommended by the engine manufacturer. During belt replacement, the crankshaft, camshaft, and idler or other sprockets should be inspected. Check the sprocket teeth for wear and damage. Check idler pulleys or sprockets for dry or loose bearings.

Many vehicles with DIS substitute a camshaft position sensor for the distributor to find the location of the #1 cylinder. This is most often used to synchronize fuel injection events to ignition events. The cam sensor is usually a magnetic hall-effect switch that gets a signal from the camshaft via a #1 aligned bump or machined surface on the cam or cam drive gear.

When inspecting camshafts, look for pitting, dull surfaces, or indications of poor lubrication. Cam or crank gears on chain drives develop wear that appears like the tooth of the chain in reverse. This condition warrants replacement. Chain deflection is the primary indicator of timing chain condition. Refer to the manufacturers' recommendations to determine chain condition. Many chain driven engines use chain guides that are tensioned by the hydraulic chain tensioner. Be sure to check for surface wear on these. If the chain has worn a groove in the guide, it will probably require replacement. Reassembly of timing components in engines with hydraulic tensioners may require that the tensioner be compressed and pinned until all components are in place.

Engines with variable valve timing will have actuation devices that allow oil flow and sensors which monitor the location of the camshaft as part of the valve train components. These items must be inspected for wear and proper operation during cylinder head repair.

Camshaft end-play is usually measured with a dial indicator. The specifications will vary depending on the size of the camshaft. A normal end-play specification can be 0.002"–0.008". Excessive camshaft end-play can be the result of a worn thrust bearing or worn thrust surfaces on the camshaft itself. Thrust bearings are replaced; if the thrust surface on the camshaft is worn, the camshaft is replaced.

14. Inspect and measure camshaft journals and lobes; measure camshaft lift.

To check camshaft straightness, rest the camshaft outer bearing journals on V-blocks and position a dial indicator against the middle cam bearing journal. Rotate the camshaft to determine runout. If the camshaft is not straight, replace it.

To measure camshaft lobes, use a micrometer to measure from the highest point on the lobe to a point on the opposite side of the lobe; record the measurement. Measure the lobe again at a position 90° from the first measurement; subtracting the second measurement from the first gives camshaft lobe lift. Replace the camshaft if lift is not within specifications.

Use a micrometer to measure the diameter of each camshaft journal. Measure this diameter in several locations. If the diameter is less than specified, replace the camshaft.

Camshaft lobe lift can also be measured with the camshaft still in the engine by removing the valve cover and mounting a dial indicator on the cylinder head. Position the dial indicator so that it contacts the pushrod tip (rocker arm removed) or rocker arm directly above the pushrod (rocker arm still in place). The dial indicator stem must be parallel with the pushrod. Crank the engine by hand and note the highest and lowest dial indicator readings. The difference between these two numbers is camshaft lobe lift.

15. Inspect and measure camshaft bore for wear, damage, out-of-round, and alignment; determine needed repairs.

Inspect the camshaft bearing bores on a camshaft-in-block for scoring or other damage. Minor nicks or burrs can be removed with emery cloth. If the bearing bores are severely damaged, the block should be replaced.

On overhead cam engines without removable bearing caps, check the bearing bores for damage. Minor nicks or burrs can be removed with emery cloth. If the camshaft is binding, or the bearing inserts are worn unevenly, use a straightedge to check bearing bore alignment. If the bores are out of alignment, the head is warped and should be straightened or replaced.

On overhead cam engines with removable camshaft bearing caps, the camshaft usually runs directly against the aluminum head. After removing the bearing caps and the cam, place a straightedge across the cam bearing surfaces to measure bearing alignment. Measure the clearance between the straightedge and each bearing bore to determine the bore alignment. When the camshaft bearing bores are improperly aligned, replace the cylinder head. Check the bearing surfaces for scoring and other damage. To measure bearing bore out of round, install the bearing caps and torque the retaining bolts to specifications. Then measure bore diameter at several locations around the bore using a telescoping gauge. If the bearing surfaces and bores are in good condition, use Plastigauge® to measure the bearing clearance.

16. Inspect valve timing; time camshaft(s) to crankshaft.

With the timing belt or chain cover removed, camshaft timing may be checked by noting the positions of marks on the camshaft and crankshaft sprockets. These marks must be aligned as indicated in the vehicle manufacturer's service manual.

On many OHV pushrod engines, the crankshaft sprocket is installed on the crankshaft nose and the crankshaft is rotated to position piston #1 at TDC. At this point, a mark stamped onto the crankshaft sprocket is pointing directly upward toward the camshaft. The camshaft sprocket is then temporarily bolted to the cam and used to rotate the cam until a mark stamped on the cam sprocket is pointing directly downward toward the crankshaft. The sprocket is then removed from the cam without allowing the cam to rotate. The timing chain is looped over the cam gear, the mark on the cam gear is positioned directly downward, and the chain is looped around the crankshaft sprocket. When the cam sprocket is attached to the cam, the timing marks on the crank and cam sprockets should be pointing toward one another.

Single overhead camshaft engines fitted with a timing belt often use a similar procedure. After positioning the crankshaft so that piston #1 is at TDC with the crank gear timing marks aligned with the timing marks on the block, the camshaft is rotated to align a mark on the cam sprocket with a mark on the cylinder head. The timing belt is then installed.

The procedure used to time camshafts on double overhead camshaft (DOHC) engines varies from manufacturer to manufacturer. On some engines the cam sprockets are friction fitted to the cams. On these engines the cams can be rotated after the timing belt is installed. When the cams are rotated to the proper positions, the bolts locking the cam sprockets to the cams are tightened. Other DOHC engines use a procedure similar to that of many SOHC engines.

Valve timing may be checked by observing the valve position in relation to the piston position. With any piston at TDC on the compression stroke, the intake and exhaust valves for that cylinder should be completely closed. When the piston is at TDC on the exhaust stroke, the intake valve should be opening and the exhaust valve should be closing. This position is called valve overlap. If the valves do not open properly in relation to the crankshaft position, the valve timing is not right. After timing component replacement, the timing mark alignment should be rechecked by turning the crankshaft two complete revolutions, ending on #1 cylinder TDC compression stroke; all timing marks should line up. If the marks fail to line up, the engine must be retimed following OEM procedures. Incorrect valve timing may cause low power or, in extreme cases, bent valves due to piston-to-valve contact.

17. Inspect cylinder head mating surface condition and finish; reassemble and install gasket(s) and cylinder head(s); replace/torque bolts according to manufacturer's procedures.

Clean and inspect the cylinder block deck in preparation for head installation. Make sure that all head positioning dowels, if used, are in place in the block. Run a thread chaser into the cylinder head bolt threaded holes. Use compressed air to eject any debris from the threaded holes. Always wear eye protection when using compressed air to clean surfaces or openings. Allowing debris or fluid to remain in threaded holes will cause false torque readings when the head bolts are tightened. Coolant or combustion leaks may result. If the holes are blind holes, fluid or debris at the bottom of the holes may cause the block to crack when the bolts are tightened.

Many newer engines are fitted with torque-to-yield (TTY) cylinder head bolts. These bolts are usually tightened to a specific torque and then rotated tighter a specified number of degrees. TTY bolts are permanently stretched as they are tightened and produce a more uniform clamping force. Most, but not all, TTY bolts must be replaced with new bolts once they are loosened. Check the manufacturer's service manual for information.

Most modern head gaskets are installed dry, without any type of sealer. When positioning a head gasket on the block, make sure that any orientation marks (up, front, left, right, etc.) are followed.

After double-checking the cylinder bores for tools, shop towels, dropped fasteners, etc., set the cylinder head on the engine block. Check the manufacturer's recommendation regarding thread lubricants or sealers. Bolts that are threaded into blind holes are often lubricated with a few drops of engine oil–some on the threads and some on the underside of the bolt head. Bolts that thread into the water jacket are often coated with a waterproof sealer.

Insert the head bolts in their holes and hand tighten them. Then tighten the bolts to specifications following the procedure and sequence specified by the engine manufacturer.

C. Engine Block Diagnosis and Repair (10 Questions)

■ 1. Remove and disassemble engine block; clean and prepare components for inspection and reassembly.

Mount the engine on a stand and remove the oil pan drain plug. Allow any oil that has accumulated in the pan during engine removal to drain into a pan. Remove the lifters from the block and store them in an organizer for later inspection. Turn the engine block upside down, remove the oil pan bolts, and remove the pan from the block. If the pan is "glued" to the block with room temperature vulcanizing (RTV) sealant, strike a strong corner of the pan with a rubber mallet to loosen it.

Check the crankshaft and connecting rod bearing caps to see if they are marked for position and direction. Main bearing caps often have numbers and arrows cast into each cap. Arrows typically point to the front (timing device end) of the engine. Connecting rods are often stamped with the cylinder number on both the rod body and cap, near the cap parting line. Take note of the direction that the numbers (or rod oil squirt holes) point for all connecting rods. If rod or main caps are not marked, mark each one with a number punch, center punch, or scratch awl.

Inspect the top of the cylinder bores for ring ridges. If the ring ridge is severe, it should be removed before attempting to remove the piston/connecting rod assemblies. Use a ridge reamer to remove the ridge (Task C.4).

Loosen the connecting rod cap bolts or nuts and remove the caps. Keep used bearing inserts with their caps for later inspection. If bolts are held captive in the connecting rod bodies, place a short length of fuel line hose over each bolt to protect the crankshaft journals during piston/rod removal.

Carefully push each piston/connecting rod assembly out of its cylinder. When an assembly is removed, immediately reinstall its mating rod cap and nuts.

Remove the harmonic balancer or pulley hub bolt from the crankshaft nose and remove the balancer or hub. Some simply slide off the nose, but most are a press fit. Use a special harmonic balancer removal/installation tool to remove a press-fit balancer. Using a jawed puller to remove a press-fit balancer will permanently damage the balancer.

Remove the timing chain/belt cover bolts and remove the cover. On engines fitted with timing chain tensioning devices, compress and lock the tensioner shoe in place, if possible. Remove the tensioner and any timing chain guides from the front of the block. Remove the oil slinger, if one is present, from the crankshaft nose.

On camshaft-in-block engines, unbolt and remove a bolted-on camshaft sprocket along with the timing chain. Remove the camshaft thrust plate bolts and the thrust plate if one is present. If the engine has a balance shaft mounted in the engine "V," remove the shaft drive mechanism.

On engines fitted with a one piece crankshaft rear main oil seal, pry the seal out of its bore or unbolt the seal mounting plate and seal from the back of the engine. Remove the crankshaft main bearing cap bolts and the bearing caps. Lift the crankshaft out of the block.

On camshaft-in-block engines, carefully withdraw the camshaft from the cylinder block, being careful to avoid nicking the bearing bores or the camshaft lobes. Use a camshaft bearing installer/remover tool to drive the bearings from their bores.

If the engine has block-mounted balance shafts, remove them now following the engine manufacturer's instructions.

Knock any core plugs loose by striking one edge with a blunt chisel. Do not strike the plug too hard and do not try to drive the plug straight into the water jacket. This could cause a bulge in a cylinder wall if the plug is driven into the wall. When the core plug tilts in its bore, grab an edge of the plug with pliers and pull the plug out of the block. Remove all oil gallery plugs.

Engine parts can be cleaned several different ways. Iron or steel parts can be soaked in a tank full of a heated alkaline solution (i.e., hot-tanked). This will remove oil, sludge, hard, baked-on carbon deposits, and mineral deposits in the coolant passages. Never put any aluminum parts in a hot tank as the caustic solution will corrode aluminum. Removal of all core and oil gallery plugs allows the chemical to be completely washed from the internal passages of the block.

Many shops have cold solvent parts washers. Small to medium size parts can be placed in the washer and sprayed with solvent. Brushes or scrapers can be used to remove stubborn deposits from the parts.

Engine parts can also be cleaned in a thermal cleaner. These are actually large ovens that heat parts to temperatures between 650 and 800°F (343 and 427°C) to oxidize the contaminants. After the thermal cleaning process, the ash is removed by shot blasting or washing the parts. The temperature inside a thermal cleaner can also be reduced to clean aluminum parts without damaging them.

Regardless of which cleaning method is used, always perform a careful inspection of oil passages in the cylinder block, head, crankshaft, and all other parts. This is especially important if the engine suffered a major failure like a spun bearing or a severely worn camshaft and lifters. Metal particles will become lodged in the oil passages and can be difficult to remove. Rod out all small diameter passages to make sure that they are unobstructed. Use a long, slender brush (called a rifle brush) to thoroughly clean oil galleries.

2. Visually inspect engine block for cracks, corrosion, the condition of passages, core and gallery plug hole condition, surface warpage, and surface finish and condition; determine necessary action.

After disassembly and thorough cleaning, you must make a careful visual inspection of the block for cracks. Any areas that are in question can be further inspected in a machine shop using dyes and MAGNAFLUX testing. Special attention to areas that may have corrosion that would affect good sealing on reassembly is important. Areas that seal coolant, such as water pump passages and core plug holes, are vulnerable. It is important to look through the core plug holes to the cylinder walls on cast iron blocks. If a lot of pitting is present, it may be necessary to replace the block to avoid internal leaks after over-boring. Be cautious of conditions like this when the cooling system condition is very poor and rusty.

Next, inspect oil galleries for sludge or casting flash that may break loose and damage the engine on reassembly. Check all gasket surfaces for pitting or warpage and make sure that they meet surface conditions set by the manufacturer. If in doubt, enlist the advice of a good machinist. Most head gaskets used in late model engines are not installed with any kind of sealer so the surface condition and cleanliness are critical to successful sealing. When checking the head surfaces or deck of the block for straightness, if you can measure any warpage with a straightedge and feeler gauge the deck should be resurfaced.

3. Inspect and repair damaged threads where allowed; install core and gallery plugs.

Inspect all threaded bolt holes in the cylinder block, especially the cylinder head bolt holes. Check holes for sludge or debris that may have been missed during cylinder block cleaning. Run the appropriate size thread chaser into each of the cylinder head bolt threaded holes to make sure that the threads are clean. Put a few drops of oil on a head bolt and thread it into each hole by hand.

If the threads in a bolt hole are damaged, the hole may be drilled to a larger size and rethreaded. A thread repair insert or HeliCoil thread insert can be installed in the oversize hole. The result is a threaded hole that is the same size as the original. Different types and brands of thread inserts are available. In most cases, however, the insert is threaded onto a special installation tool, coated with thread locking compound, and then threaded into the oversize hole. The installation tool is then removed and a hammer and punch are used to break off a tang at the bottom of the insert.

Inspect oil gallery plug threaded holes for dirt or damage. Do not run a tap very far into these holes since they usually have tapered pipe threads. Run a rifle brush down the oil galleries to make sure that all debris has been removed. Coat the new oil gallery plugs with teflon tape or an oil-resistant sealer and thread them into the block. Do not overtighten the plugs. If there are small core plugs at the ends of the oil galleries, coat the edges of new plugs with an oil-resistant sealer and then drive the plugs into the block. Use a cold chisel and hammer to cross stake the end of the bores after the core plugs are installed.

Clean out core plug bores with emery cloth before installing new plugs. If a bore is damaged, it may be repaired by boring it to the next specified oversized plug. Oversized core plugs are stamped with the letters OS. Before installing a new plug, coat the sealing edge with a nonhardening, water-resistant sealer. Drive the plugs into the block using the proper special driving tool. Make sure that the plug goes into the bore squarely to prevent leaks.

4. Inspect and measure cylinder walls; hone and clean cylinder walls; determine need for further action.

Use a dial bore gauge to measure the cylinder diameter in three vertical locations. These locations are just below the ring ridge at the top of the cylinder, in the center of the ring travel, and just above the lowest part of the ring travel. Cylinder taper is the difference in the cylinder diameter at the top of the ring travel compared to the diameter at the bottom of the ring travel.

In each of the three vertical cylinder measurement locations, measure the cylinder diameter in the thrust direction and in the axial direction. Cylinder out of round is the difference between the cylinder diameter in the thrust and axial directions. If cylinder out of round exceeds specifications, rebore the cylinder.

If cylinder wear, out of round, and taper are within specifications, the cylinders may be deglazed. Very mildly worn cylinders should be deglazed with a brush hone, which removes material very slowly. Moderately worn cylinders (still within specifications) may be deglazed with 220 or 280 grit stones installed on a cylinder hone. When the honing operation is completed, the cylinders should have a 50° to 60° crosshatch pattern. After deglazing, the cylinder should be cleaned with hot, soapy water and a stiff bristle brush. Ordinary solvent will not remove grit from pores in the cylinder wall–use hot soapy water. The bores are clean when a clean, lint-free cloth is used to wipe them and the cloth does not get dirty. When the bores are clean, rinse the block and dry it thoroughly. Coat all machined surfaces with a light coating of the manufacturer's recommended engine oil.

If one cylinder requires reboring, most manufacturers recommend reboring all the cylinders to the same size. Cylinder reboring usually is done with a specialized piece of equipment called a boring bar. Cylinders must be honed after boring. A honing machine is usually used for this. After cylinder honing, the same procedure for block cleaning should be followed as previously discussed in cylinder deglazing.

▇ 5. Inspect crankshaft for end-play, journal damage, keyway damage, thrust flange and sealing surface condition, and visual surface cracks; check oil passage condition; measure journal wear; check crankshaft reluctor ring/tone wheel (where applicable); determine necessary action.

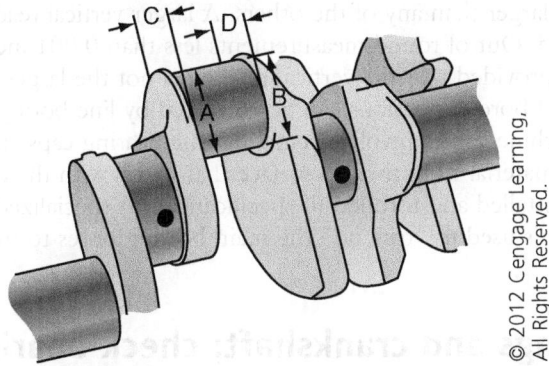

© 2012 Cengage Learning,
All Rights Reserved.

Crankshafts must be inspected at several different points to ensure dependable engine performance.

End-play is the amount the crank moves fore and aft in the block. One of the crank shaft bearings provides a thrust surface that controls fore and aft movement. Crank wear or bearing wear can allow excessive end-play which can affect clutch operation in manual transmission equipped vehicles, belt alignment problems, starter engagement problems, and even automatic transmission pump or converter damage. This measurement is made when the crank and main bearings are assembled in the block by using a dial indicator in a horizontal or parallel mount to the crank and measuring total fore and aft movement.

A crank can be measured for straightness by placing it in a set of V-blocks and using a dial indicator to measure the runout of the main bearing surfaces. Another good hint of a bent crank is excessive main bearing wear on one or two journals compared to others. This is not an indicator by itself as each journal's bearing clearances can be a factor as well.

The crankshaft bearing surfaces must be inspected for cracks and wear grooves. The crank must be replaced, resurfaced, and/or repaired if problems are found here. Clean out the oil passages with small gallery brushes to be sure no restrictions are present.

Using the diagram provided, we need to make some measurements to determine if it is necessary to resurface the crank. Using a micrometer, we need to determine the difference between point A and B. This will tell us if the bearing surface is tapered. Comparing A to C and B to D will tell us if we have an out of round condition. If the crank is even, you can determine bearing clearance with a standard bearing by comparing your measurements to the specs for the crank. If the crank is not larger than the minimum specification, it may be necessary to resurface the crank to the next undersize and use appropriate bearings.

6. Inspect and measure main bearing bores and cap alignment; mark caps for location and direction.

Check main bearing bore alignment before bearing caps are installed. The block should be resting on a flat surface, not hanging from an engine stand by the flywheel end. Lay a straightedge across the bores and check alignment using feeler gauges. If bearing bores are not aligned, the block can be line bored.

Check the bearing cap and cylinder block mating surfaces for nicks and burrs. These can be removed using a file. Make sure that the caps are marked for proper location and direction of installation. If the caps are switched and the correct position cannot be positively determined, the block will have to be line bored.

When measuring main bearing bore diameters, the bearing caps must be installed and the bolts properly torqued. Check bore diameter in three directions. The vertical measurement should not be larger than any of the others. A larger vertical reading indicates the bore is stretched. Out of round measurements less than 0.001 inch (0.025 mm) are acceptable, provided that the vertical reading is not the largest. Improper bore alignment and bore dimensions can be corrected by line boring. This operation, performed by machine shops, involves removing the bearing caps and planning a small amount of material from the cap surface that mates with the cylinder block. The caps are then reinstalled and torqued to specifications. A specialized piece of equipment called a line hone is used to "true up" the main bearing bores to their original diameters.

7. Install main bearings and crankshaft; check bearing clearances and end-play; replace/torque bolts according to manufacturers' procedures.

Clean the main bearing bores in the cylinder block and bearing caps with solvent and allow the bore surfaces to dry. Do not oil the bores. Handle new bearing inserts being careful to avoid touching the bearing surface with your fingers. Wipe the back of the bearing inserts with a solvent-dampened cloth and allow the inserts to dry. Install the bearing inserts in the cylinder block and main bearing cap bores. The upper bearing halves are usually grooved and each contains an oil supply hole. Make sure that the oil hole in the bearing aligns with the oil hole in the bearing bore. Make sure that the tab on each bearing insert fits tightly in its bearing bore notch.

Carefully lay the crankshaft in the cylinder block.

To measure bearing clearance, install a strip of Plastigauge across each journal. Install the bearing caps and bolts, tightening the bearing cap bolts to the specified torque. Remove the bearing cap bolts and the bearing caps. Compare the width of the crushed Plastigauge strip on the bearing journal to the scale provided on the Plastigauge package to determine the bearing clearance.

Crankshaft end-play may be measured by inserting a feeler gauge between the crankshaft thrust journal and the thrust lip on one of the main bearings. On some engines, a dial indicator is used to measure crankshaft end-play while moving the crankshaft with a pry bar. Excessive end-play may cause premature bearing wear or noise as the crank chucks back and forth in the block.

If bearing clearance and end-play are within specifications, remove the main bearing caps and the crankshaft. If a two piece rear main seal is used, install a rope-type or split lip seal in its grooves in the block and rear main bearing cap. Oil the main bearing inserts, lay the crankshaft in the block, and install the main bearing caps.

Install the bearing cap bolts and torque them to specifications. Some engines are fitted with TTY main bearing cap bolts that must be replaced after being loosened. Check the manufacturer's service manual to determine whether the engine you are servicing has TTY main bearing cap bolts.

8. Inspect camshaft bearings for excessive wear and alignment; replace bearings if necessary; install camshaft, timing chain, and gears; check end-play.

Inspect the camshaft bearings for scoring, roughness, and wear. Camshaft bearings or bearing bores should be measured at two different locations with a telescoping gauge. Measure the camshaft journals with a micrometer, and subtract the journal diameter from the bearing diameter to obtain the clearance. If the wear exceeds specifications, replace the bearings.

The type of tool needed to remove and install the camshaft bearings depends upon the engine design. Most overhead valve (OHV) camshaft-in-block engines will use a camshaft bushing driver and hammer. The right size mandrel is selected to fit the bearing. Turning the handle tightens the mandrel against the camshaft bearing and then the bearing is driven out by hammer blows. The same tool is used to replace the bearings. Some OHC engines require a special puller/installer.

Never attempt to remove and install camshaft bearings in an OHC cylinder head using a bushing driver. The bearing supports may be bent or broken due to the hammer blows, especially on aluminum heads.

When installing cam bearings, it is very important that the bearing insert be properly positioned in the bearing bore. Be absolutely sure that any oil hole(s) in the bearing insert align with oil supply passages in the bearing bore. This may mean that the insert is positioned toward the front of the bore, the back of the bore, or even in the center of the bore. The position is not important as long as the oil holes line up.

Many overhead cam engines do not have removable camshaft bearings–the camshaft journals run directly against bearing bores machined into the aluminum cylinder head. Inspect the bearing surfaces on these engines for scoring, roughness, and wear. Measure the bearing bores at two different locations with a telescoping gauge. Measure the camshaft journals with a micrometer, and subtract the journal diameter from the bearing diameter to obtain the clearance. If the wear exceeds specifications, replace the cylinder head.

When the camshaft bearing bores are machined into the cylinder head, the bearing caps should be removed and a straightedge positioned across all the bearing bores. Insert a feeler gauge between the straightedge and each bearing bore to measure any misalignment. Misalignment indicates that the cylinder head may be warped. Have a machine shop check for this. Cylinder heads can sometimes be straightened. Severely warped heads should be replaced.

Camshaft end-play is normally checked using a dial indicator. A typical specification will be 0.002"–0.008". End-play which is greater than specification can be caused by a worn camshaft thrust washer or worn thrust surfaces on the camshaft. If either the thrust washer or the cam is worn they are replaced. Camshaft end-play which is less than specification is generally caused by a failure of the technician to reassemble the components correctly. Excessive camshaft end-play can result in camshaft position sensor damage.

9. Inspect auxiliary (balance, intermediate, idler, counterbalance, or silencer) shaft(s), drive(s)/gear(s), and support bearings for damage and wear; time balance shaft to crankshaft; determine necessary action.

Balance shafts are found on many 4- and 6-cylinder engines. Some rotate at crankshaft speed, while others rotate at twice crankshaft speed. Removal and installation procedures can vary widely so refer to the manufacturer's service manual for service information.

Some balance shafts are mounted to the bottom of the cylinder block and are chain driven off the crankshaft. Some are gear driven by a large gear machined into a disc that is part of the crankshaft. Belt-driven balance shafts are often mounted inside the bottom of the cylinder block, much like camshafts on camshaft-in-block engines. On some popular V-type engines, a balance shaft is mounted in the cylinder block directly above the camshaft and gear driven off the camshaft.

Once removed, balance shafts should be checked for runout with the same procedure used for measuring camshaft runout. The balance shaft journals should be measured for taper with the same procedure for measuring crankshaft journal taper. When the balance shafts are installed, they must be properly timed to the crankshaft or severe engine vibration may occur upon engine startup.

10. Inspect, measure, service, or replace pistons and piston/wrist pins; identify piston and bearing wear patterns caused by connecting rod alignment problems; determine necessary action.

Inspect pistons for cracks and damage from overheating. Pistons with these conditions should be replaced. Inspect the piston skirts for uneven wear and scoring. Wear on the edges of the piston skirt next to the wrist pin hole may be caused by a bent or twisted connecting rod.

Clean the piston ring grooves using a ring groove cleaning tool. Be careful to avoid removing material from the bottom of the groove. Make sure that oil drain holes at the bottom of the oil ring groove are not obstructed.

To measure piston ring side clearance, insert a new ring in the piston groove and position a feeler gauge between the ring and the groove. If ring side clearance is excessive, the piston should be replaced.

If the piston passes the inspections mentioned previously, check for a worn wrist pin bushing. Clamp the connecting rod body lightly in a vise (use soft jaw covers) and attempt to rock the piston against the connecting rod sideways at a right angle to normal piston/ connecting rod motion. If any play is noticeable, the piston wrist pin bore, wrist pin, or connecting rod small end bushings are worn and the piston and connecting rod must be separated. If no play is noticeable and the piston is to be reused, the components need not be separated.

Pistons must be fitted to their cylinder bores. If piston-to-cylinder wall clearance is excessive, piston slap may be noticeable. If there is not enough clearance, piston scuffing will occur. Check piston clearance by measuring the cylinder bores and the piston diameters.

To measure piston diameter, position a micrometer to contact the piston thrust surfaces (the surfaces at right angles to the piston, or wrist, pin bore). The exact measuring location varies according to manufacturer. Most manufacturers, however, specify a point about $3/4$ inch below the wrist pin centerline.

Bearing wear patterns can help the technician identify worn or damaged components. Bearings that are worn at opposite corners are an indication of bent or twisted connecting rods. Bent or twisted connecting rods can be caused by liquid on top of the piston resulting in a hydraulic lock. Wear at the outer edges of the bearing indicates radius ride. Radius ride generally occurs when the fillet radius was not correctly machined into the crankshaft. Normal rod bearing wear will have the upper half of the rod bearing wearing faster than the lower half. However, if the upper bearing is greatly worn and the lower bearing shows little wear, then "lugging" of the engine should be suspected.

11. Inspect connecting rods for damage, bore condition, and pin fit; determine necessary action.

Inspect the connecting rods for cracks and obvious damage. Remove the cap nuts, caps, and bearing inserts. Check that the cap bolts are not loose in the rod body.

Inspect the bearing inserts for uneven wear. If the front and rear edges of a bearing are worn more than the center area, the rod may be bent or twisted. If the bearing inserts are worn more at the parting line areas, the rod big end bore may be stretched. Machine shops have special jigs used to check for bent/ twisted connecting rods. If rod bend or twist exceeds specifications, replace the connecting rod.

Measure the connecting rod big end bore for taper, out of round, and proper bore size. If any of these dimensions are not within specifications, the connecting rod should be reconditioned or replaced.

On piston/connecting rod assemblies with free-floating wrist pins, remove the pin retainer circlips or snap rings and slide the wrist pin out of its bores. Take note of piston to rod orientation and separate the piston and rod. On piston/connecting rod assemblies with press-fit wrist pins, use the appropriate press and adapters to remove the wrist pin. Measure the connecting rod small end bore (or "eye") diameter. If bore diameter exceeds specifications, it may be possible to ream out the bore and install an oversize wrist pin. Some rod eyes have a pressed in bushing. If the bore is worn on this type of rod a new bushing can be installed. The new bushing must be reamed to fit the wrist pin.

If the rod beam has an oil squirt hole, make sure that the passage from the hole to the big end bore is not obstructed.

12. Inspect, measure, and install or replace piston rings; assemble piston and connecting rod; install piston/rod assembly; check bearing clearance and side-play; install connecting rod bearings; replace/torque fasteners according to manufacturers' procedures.

Before new piston rings are installed, ring end gap must be checked and adjusted. Compress a ring just enough so it fits in a cylinder. Insert a piston in the cylinder upside down (crown end first) and push the ring down near the bottom of the cylinder. The ring should be about 0.5 inch from the bottom. Use a feeler gauge to measure the gap where the piston ring ends meet. If the gap is too small, the ring ends must be filed to increase the gap. If the gap is too large, the wrong rings were selected or the cylinder was bored/honed incorrectly.

If the pistons and connecting rods were separated, reassemble them. Position the rod eye inside the piston, making sure that the parts are oriented correctly. Most pistons have a notch or arrow in their crown that must point toward the front of the engine. Connecting rod orientation varies according to the engine manufacturer. Refer to notes made during disassembly or check the service manual for information. On engines with free-floating wrist pins, dip the pin in engine oil and install it in its bore. Install new circlips or snaprings. When installing press-fit wrist pins, many manufacturers recommend that the rod eye be heated before pin installation. Some manufacturers recommend that the piston be heated in a piston heater as well. When components are heated as necessary, use a press and the appropriate adapters to install the pin.

When installing the piston rings on the piston, install the oil rings first. Follow instructions that come with the rings to position the upper and lower rail gaps with respect to the expander gap. Oil ring rails can be spiraled into their slots. When installing compression rings, make sure that any marks stamped into the ring are facing upward. Install the bottom compression ring first and then the top compression ring using a piston ring expander to prevent ring distortion or breakage. Do not spiral compression rings onto the piston. This can distort the rings, causing them to resemble a lock washer.

Clean the bearing bores in the connecting rods and rod caps and allow the bore surfaces to dry. Handle new bearing inserts being careful to avoid touching the bearing surface with your fingers. Install the bearing inserts in the rod and cap bores. If the rod has an oil squirt hole, make sure that the oil hole in the bearing aligns with the oil hole in the rod body. Make sure that the tab on each bearing insert fits tightly in its bearing bore notch.

Before installing the piston/connecting rod assembly in the block, install short lengths of fuel line, or "chopsticks," over the rod bolts and position the piston ring end gaps according to the manufacturer's instructions. If no instructions are given, it is common practice to position the gaps as follows:

1. Oil ring expander gap facing the front of the engine, directly above the wrist pin centerline.
2. Upper oil ring rail gap 45 degrees to one side of the expander gap.
3. Lower oil ring rail gap 45 degrees to the other side of the expander gap.
4. Bottom compression ring gap on the left side of the piston (90 degrees from the wrist pin).

5. Top compression ring gap on the right side of the piston (90 degrees from the wrist pin).

Rotate the crankshaft to position the crank pin for the piston/rod being installed at bottom dead center (BDC). Lubricate the piston, rings, wrist pin, and cylinder bore with clean motor oil. Install a ring compressor on the piston and slide the piston/rod into the block. Push the rod body against the crank pin, remove the rubber protective hose, and temporarily install the rod cap and nuts. Install the remaining piston/rod assemblies.

To measure bearing clearance, remove the rod cap and install a strip of Plastigauge across the crankshaft journal. Install the bearing cap and nuts, tightening the nuts to the specified torque. Remove the bearing cap nuts and the bearing caps. Compare the width of the crushed Plastigauge strip on the bearing journal to the scale provided on the Plastigauge package to determine the bearing clearance. After removing the Plastigauge, lubricate the connecting rod bearing and crankshaft journal with engine oil, install the cap, and torque the fasteners to specifications. It is a good practice to rotate the crankshaft a few times after the installation of each piston and rod assembly. If there is any binding, you will know immediately which cylinder is affected.

Measure the side clearance at each connecting rod by inserting a feeler gauge between the side of the connecting rod and the edge of the crankshaft journal. Alternately, some manufacturers will use a dial indicator to perform this measurement. If side clearance exceeds specifications, the sides of the connecting rod or crankshaft journal is worn. Excess side clearance results in an excessive amount of oil on the cylinder walls and can contribute to oil consumption.

13. Inspect, reinstall, or replace crankshaft vibration damper (harmonic balancer).

Special tools are required to remove and install the vibration damper. The use of a regular outside gear puller to remove the vibration damper could damage the damper.

Inspect the rubber between the inner hub and outer inertia ring on the vibration damper. If this rubber is cracked, oil-soaked, deteriorated, or protruding from the damper, replace the damper. If the inertia ring on the damper is loose or has shifted forward or rearward on the hub, replace the damper. The inertia ring may also have rotated on the hub and the ignition timing reference would be incorrect. Inspect the vibration damper hub for cracks or a damaged keyway. If either of these conditions is present, replace the damper. Inspect the seal contact area on the hub for a wear groove or scoring. If either of these conditions is present, replace the damper or install a sleeve on the hub to provide a new seal contact area.

14. Inspect crankshaft flange and flywheel mating surfaces; inspect and replace crankshaft pilot bearing/bushing (if applicable); inspect flywheel/flexplate for cracks and wear (includes flywheel ring gear); measure flywheel runout; determine necessary action.

Inspect the crankshaft flange and the flywheel-to-crankshaft mating surface for metal burrs. Remove any metal burrs with fine emery paper. Be sure the threads in the crankshaft flange are in satisfactory condition. Replace the flywheel bolts and retainer (if fitted) if any damage is visible on these components. Install the flywheel, retainer, and bolts, and tighten the bolts following the torque and sequence provided by the engine manufacturer.

Inspect the flywheel for scoring and cracks in the clutch contact area. Minor score marks and ridges may be removed by resurfacing the flywheel. If deep cracks or grooves are present, the flywheel should be replaced.

Mount a dial indicator on the engine block or flywheel housing and position the dial indicator stem against the clutch contact area on the flywheel. Rotate the flywheel to measure the flywheel runout. If runout exceeds specifications, replace the flywheel.

Insert a finger in the inner pilot bearing race and rotate the race. If the bearing feels rough or loose, replace the bearing. Check the pilot bushing to verify that it is not loose. A transmission input shaft may be positioned in the pilot bushing to check for excessive play. If too much play exists, replace the bushing. A special puller may be used to remove the pilot bearing or bushing. The proper driver must be used to install the pilot bearing or bushing. Always verify that the transmission input shaft fits in a new bushing before attempting to install the engine (or transmission).

Inspect the starter ring gear for excessive wear or damage. On manual transmission flywheels, the starter ring gear is often replaceable. Remove the old gear by drilling a hole through the gear at the "root" between two teeth. Position a cold chisel between the two teeth and strike it with a hammer. Take note of whether the gear has a chamfer on one side before removing it. To install the new gear, first heat it to about 400°F in an oven, and then slip it over the flywheel body and allow it to cool.

15. Inspect and replace pans and covers.

Always inspect the gasket mounting surfaces on sheet metal pans for warpage and look for dished retaining bolt holes. Dished mounting holes and warped mounting surfaces must be straightened by hammering them flat again. Use a straightedge to verify that gasket surfaces are flat.

Gaskets are used to seal minor variations between two flat surfaces. Oil pan gaskets or rocker arm cover gaskets are usually manufactured from cork, rubber, or a combination of rubber and cork or rubber and silicone. Timing cover, water pump, and thermostat housing gaskets on older engines were made of specially treated paper. On newer engines, these same components are sealed to the block using synthetic rubber o-rings or silicone "spaghetti" seals. Some paper gaskets may require the use of a nonhardening sealer; always follow OEM and gasket manufacturer's guidelines for sealer usage on gaskets. Synthetic rubber and silicone seals are often installed without coatings or cements.

16. Assemble the engine using gaskets, seals, and formed-in-place (tube-applied) sealants and thread sealers according to manufacturer's specifications; reinstall engine.

Gasket sealing technology has changed dramatically in a relatively short period of time. This requires the technician to have a large base of knowledge on how to seal the many types of gaskets. The newest types of gaskets are the captive silicone or rubber gaskets that fit into a channel cut or cast into one of the components' mating surfaces. These gaskets are very reliable and depend on surface tension and crush to seal. Many of these gaskets are reusable in low-pressure environments. Reusable gaskets for areas such as oil pans are a hybrid version of this type of gasket, where the gasket is captive in a high density plastic and TTY bolts are used to tension the component. These gaskets are all installed dry on very clean surfaces. RTV sealer is used in some applications to actually replace the gasket altogether. In these situations, the application is critical. Any air bubbles or gaps in the sealer will develop into leaks later on. This type of sealer comes in many varieties. In vehicles with computer controls, the use of a sealer complimentary to oxygen sensors is important. The gases released during the vulcanizing process, and for some time after, will cause the oxygen sensor to fail. Another gasketless environment is the anaerobic sealer that dries when oxygen is not present. This is used as a sealer in applications where two machined mating surfaces must seal and maintain a precision location or clearance when assembled.

Seals are usually in the form of lip seals that fit over shafts allowing the sealed component to move. These seals depend on the oil they seal to provide lubrication. Due to this fact, these seals have the highest failure rate. A rear main seal is an example of this type of seal whether it be full circle or split type. The surfaces these seals mate with are subject to wear that leaves a groove over time. These grooves can cause repeated replacements if they are not removed, the component replaced, or a repair sleeve installed. When installing a seal, it is recommended to lightly lubricate the shaft and the seal surface that contacts the shaft prior to installation; this will help prevent premature seal failure.

In all cases where sealing is the goal, cleanliness is critical to success. It is important to use cleaners that do not leave residue behind. Chlorinated cleaners or denatured alcohol are the best choices.

In many applications, a thread sealer is called for to control coolant or oil where bolts pass into areas containing fluids. It is important to use the sealer designed for the application as some sealers may be oil or coolant soluble.

When reinstalling the engine, care should be taken to prevent the trapping of the wiring harness between the engine and transmission and/or the engine and mounts. Wiring harnesses and hoses should have retaining clips reinstalled in the proper location. Failure to secure these items properly can result in abrasive wear and damage.

D. Lubrication and Cooling Systems Diagnosis and Repair (8 Questions)

1. Diagnose engine lubrication system problems; perform oil pressure tests; determine necessary action.

In most cases, diagnosis of engine lubrication problems is more forensic than preventive. Oil is critical to lubrication as well as internal engine cooling. When a component does not receive adequate oil, it will fail quickly. The most common cause of oil related failures is plain old dirty oil. When oil breaks down, it can begin to adhere to internal engine surfaces creating sludge or a varnish that can restrict oil passages or drains. Lifters and valve lash adjusters are particularly sensitive to these problems due to very small passages inside them. Something as simple as a stuck open thermostat can cause the engine to not reach operating temperature, promoting oil sludging. This sludge can find its way back into the oil pan and restrict the oil pump pick-up tube.

On the diagnostic side of the oil system, technicians are often called on to determine the cause of low oil pressure readings. This calls for removing the oil pressure sender or switch, and checking oil pressure with a mechanical gauge. Manufacturers have their own specs and different RPM readings that they want tests to be conducted at, so do not assume that 20 psi is low unless the manufacturer tells you so. In the case of complete loss of oil pressure, there are a couple of common causes. The pressure relief valve in the oil pump can stick open causing by-pass within the pump. The other common cause is blockage of the oil pump pick-up with sludge. Both are the result of poor maintenance. Low oil pressure can be caused by the aforementioned causes, excessive bearing clearances (high mileage engines fall victim to this), an oil filter that was installed too tight, or some other internal leakage. High oil pressure is usually only caused by a stuck closed pressure relief valve, higher viscosity oil than called for, or installing the incorrect oil pump. This will usually only occur in a situation where a high volume or high pressure pump was substituted for the original design.

If all pressures meet specs and the gauge was reading out of spec, you must consider and test the electrical components or the oil pressure warning system.

2. Disassemble and inspect oil pump (includes gears, rotors, housing, and pick-up assembly); measure oil pump clearance; inspect pressure relief devices and pump drive; determine necessary action.

Inspect the oil pump pressure relief valve for sticking and wear. If this valve sticks in the closed position, oil pressure will be too high. A pressure relief valve stuck in the open position results in low oil pressure. A weak pressure relief spring will result in excessive return of oil to the pan and possible low oil pressure or volume.

On rotor type oil pumps, measure the thickness of the inner and outer rotors with a micrometer. When this thickness is less than specified on either rotor, replace the rotors or the oil pump. The following oil pump measurements should be performed with a feeler gauge:

1. Measure pump cover flatness with a feeler gauge positioned between a straightedge and the cover.
2. Measure the clearance between the outer rotor and the housing.
3. Measure the clearance between the inner and outer rotors with the rotors installed.
4. Measure the clearance between the top of the rotors and a straightedge positioned across the top of the oil pump.

On gear type oil pumps, inspect the gears and housing for scoring and excessive wear. When reassembling a gear type pump, be sure to align any match marks stamped onto the pump gears.

Oil pump pick-up tubes should be carefully cleaned or preferably replaced if any sludge is present. While it would seem that oil filter replacement is a no-brainer, do not forget to rule out that a filter installed too tight can cause oil delivery problems.

3. Inspect, test, and replace internal and external engine oil coolers.

The engine oil cooler helps to remove heat from the engine by removing heat from the lubricating oil. There are many different variations of these coolers. Some are externally mounted oil-to-air coolers; these are generally mounted in front of the radiator and transfer the lubricating oil heat directly to the air. Other styles will transfer the heat of the lubricating oil to the engine coolant, and the coolant will transfer the heat to the air. This second style can be located in an engine block cavity, incorporated into the oil filter mounting base, or be incorporated into the radiator. If the oil-to-air cooler leaks, the leak will be external and generally easy to diagnose. If the oil-to-coolant style of cooler leaks, there will be a mixing of oil and coolant.

There are many methods to check for oil in coolant such as looking for a milky grey sludge in the lube system, or looking for a skim of oil in the coolant. If coolant in the oil is suspected, let the engine set overnight, then loosen the oil drain plug. If the first drop to appear is coolant-colored (green, red, orange etc.), there is coolant in the oil. If the first drop is clear water, the suspected contamination is not coolant but water. If the first drop is oil, there is no contamination. Oil coolers can be pressurized with shop air and submerged in hot water to check for leaks. Normally, leaking oil coolers are replaced not repaired.

Complete servicing of the oil cooler should be done as part of a major engine overhaul, or whenever there is oil in the water, water in the oil pan, or a milky gray sludge in the filter. In all of these cases, specific manufacturer's instructions will apply. Look for places where the oil cooler baffles may have vibrated against the tubes. This common failure results in leaks. If you find signs of this, replace the cooler. Look for gasket particles and debris that is obstructing fluid flow and clean thoroughly. If the engine has suffered a major bearing failure, the oil cooler will be filled with metal particles. In this case, cleaning the cooler should not be attempted, it should be replaced. Check the lines and hoses for signs of leaking. Use a small light to inspect the inside of hoses, since they deteriorate from the inside first. Replace any hoses with internal cracks.

4. Fill crankcase with oil and install engine oil filter.

When changing the oil and filter, attention to detail is important. It cannot be overemphasized to use the correct viscosity engine oil. Some manufacturers now require very specific engine oils, such as 0W-20, or specialized synthetics manufactured to such exacting tolerances that there will be only one or two oils marketed to meet those specifications. In some instances, the oil will only be available from the dealer. Using the incorrect oil can void the warranty and result in engine damage. Engine oil should only be poured from clean, sealed containers. If the oil is stored in bulk containers and transport cans are used to move the oil to the engine, extra care must be used to ensure that these items are kept clean.

Oil filters have come in the spin-on replaceable canister-style for many years. In this style, the filter is permanently installed in a canister and the technician simply screws the old one off, cleans the seal contact area on the engine, wipes clean oil on the new seal, reinstalls the filter, and tightens $3/4$ of a turn after the seal contacts the base. Many manufacturers have now moved to the replaceable cartridge-type filter similar to what was used prior to 1965 on most engines. In this system, the filter element is removable from the housing. The only items discarded during the oil change are the filter cartridge itself and the sealing o-rings. This style of filter produces less waste; however, it does complicate oil filter servicing. Since the new filter itself is exposed to dirt during handling, care must be taken to ensure that the new filter element is not contaminated. The new o-rings must be installed in the appropriate location and lubricated. The filter will have a torque specification which must be followed.

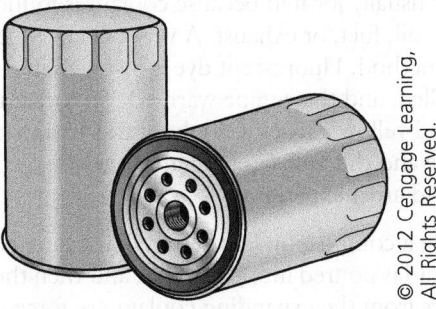

© 2012 Cengage Learning, All Rights Reserved.

Oil pans can be stamped style or aluminum. When reinstalling the oil drain plug, it is very important to use a torque wrench. Over-torqueing the oil drain plug is a common mistake. Eventually this will result in damaged threads in the oil pan which will require repair. Oil drain plugs are typically of the straight thread O-ring style, which will use a copper washer or o-ring to obtain a seal. This sealing washer, or o-ring, will need to be replaced at each oil change.

Recommended engine oil change intervals vary widely. Although most technicians will recommend 3,000 mile oil changes, factory oil life monitors may extend this to as much as 10,000 miles. When making recommendations to the customer, care must be taken to explain the advantages and disadvantages of extended oil change intervals.

When installing oil in a new or overhauled engine, the lubrication system must be primed. This can be accomplished on some engines by spinning the oil pump with a drill; however, this is becoming less likely due to the fact that few new engines now have an oil pump driven by the distributor. Another method is to keep the engine from starting by preventing fuel delivery and cranking the engine with the starter motor until oil pressure is indicated on the gauge. Yet another method is to use a specialized lubrication system priming tool which can be contacted to an oil passage in the engine, and pressurized engine oil forced through the system before the engine is started.

5. Perform cooling system pressure tests; perform coolant dye test; determine necessary action.

Whenever it is necessary to replenish coolant frequently, a pressure test should be performed to locate the source of the loss. Pressure testing will easily locate most external leaks. Leaks are usually located by a good visual inspection, since antifreeze leaves a film wherever there is a leak. Cold leaks occur due to contraction of sealing components, especially hose clamps. They often cease to leak at operating temperatures. Silicone hoses are frequently used due to their longer service life. Use the special clamps they require, and be careful to torque them properly since they are sensitive to over-tightening. A cooling system pressure testing kit includes a hand-actuated pump and gauge, and the necessary adapters to fit different radiators and pressure caps. Attach the pump using the necessary adapters to the radiator and pump up pressure to the pressure range of the cooling system. Observe the system's ability to hold pressure while looking for visible signs of leakage.

The radiator cap may also be tested using the pump and gauge assembly. Radiator caps are rated by the pressure to overcome the caps, spring pressure and unseat the seal. When this occurs, coolant is routed to the surge tank. When the cooling system cools after the engine is shut down, pressure drops. When it falls to a point where atmospheric pressure is about 0.25 psi higher than the coolant pressure, a vacuum valve in the cap unseats and coolant is forced by atmospheric pressure from the surge tank back to the radiator. When this valve fails to open, radiator hoses may be "sucked" flat.

Internal leaks are harder to find and are usually located because coolant is found in places where it does not belong, such as in the oil, fuel, or exhaust. A very effective method to locate internal coolant leaks is the dye method. Fluorescent dye is installed in a cold cooling system, the radiator cap reinstalled, and the engine warmed. To locate the leak, the technician then uses a black light and yellow glasses to look for the area that glows. Originally, the black light was expensive and the bulbs in the light were fragile; now the lights are made much like a flashlight and are inexpensive.

Yet another coolant test method is using a combustion leak tester. This tester is installed at the radiator filler neck, a blue test liquid is poured into the tester, and then the engine is started. As the engine warms, the vapors from the expanding coolant are forced through the tester. The liquid will change from the blue to yellow if the vapors contain combustion chamber gases.

6. Inspect and test radiator, heater core, pressure cap, and coolant recovery system; replace as required.

Examine the radiator for obvious damage or defects. Look for bent fins and fins clogged with dirt, road debris, or insects. These conditions greatly reduce radiator efficiency and can cause engine overheating. If damage is not severe, bent fins can usually be straightened using a special comb made for this purpose. Dirt and insects can be removed using a stream of low-pressure water or compressed air.

Check the radiator for leaks or damp spots. Cracked solder seams and corroded tubes in copper/brass radiators can allow coolant to leak very slowly without leaving puddles of coolant. The same is true for aluminum/plastic radiators with cracked plastic tanks and leaking tank gaskets. Hard to find leaks can be located by removing the radiator, plugging the inlet and outlet fittings, and pressurizing the radiator with a cooling system tester. Submerge the radiator in a tank of water and check for bubbles.

Check the radiator cap for corrosion and damaged or deteriorated gaskets; also check the radiator filler neck seat. If the cap gasket or filler neck seat is damaged, the cooling system may not pressurize enough to prevent boilover. Coolant will be forced out of the cooling system and onto the ground or into the coolant recovery tank, if the vehicle has one. The engine may overheat.

If the vehicle is equipped with a coolant recovery system, check the gasket at the very top of the radiator cap. If this gasket is missing or leaking, coolant may be forced into the recovery tank when the engine warms up, but may not be drawn back into the radiator during cool down. Check the tube that connects the radiator filler neck to the reservoir for kinks, damage, or loose connections. These conditions may also allow coolant to flow into the tank, but not return to the radiator. Check the coolant reservoir for cracks, loose fittings, and other damage. Some recovery systems use a reservoir cap that allows a length of tubing to hang down into the coolant. If this hose is missing or damaged, coolant cannot return to the radiator when the engine cools down.

7. Inspect, replace, and adjust drive belt(s), tensioner(s), and pulleys.

Inspect the accessory drive belts for condition and tension. On conventional V-belts, the sides of the belt are the friction surfaces, so check the sides for cracks, glazing, or loose cord material. Replace a belt showing any of these conditions. If a V-belt is severely worn, it may contact the bottom of the pulleys. Replace severely worn belts. If a belt is severely worn on just one side, check pulley alignment. V-belt pulleys must be aligned within $1/16$ inch (1.6 mm) per foot of belt span. If pulleys are not aligned, check for loose accessory mounting bolts, missing spacers, or bent brackets.

V-belt tension can be checked using a variety of special testers. With one type of tester, the tool is placed over the belt at the center of a belt span. Squeezing the tool handles causes the tool dial to display belt tension, usually in pounds. Belt tension can also be checked by measuring the amount of belt deflection with a ruler. Use your thumb to press on the belt at the middle of a span while holding the ruler next to the belt. If belt tension is correct, the belt should deflect 0.5 inch (12.7 mm) for every foot (30.5 cm) of belt span.

A moderately loose or worn belt may cause a squealing noise when the engine is accelerated. A severely worn or loose belt may cause a discharged battery, engine overheating, or a lack of power steering assist. An overtightened belt may fail suddenly, or damage the alternator front bearing. An overtightened belt can also cause the upper half of the crankshaft front main bearing to wear prematurely.

When repositioning an accessory device (alternator, power steering pump, etc.) to adjust belt tension, always look for the pry points provided by the manufacturer. Some devices have slots for inserting a large screwdriver or pry bar. Others have built-in square holes to accommodate a 0.5 inch breaker bar or ratchet. Never pry on a power steering pump housing to tighten the drive belt. These housings are not meant to withstand such abuse and will be damaged, possibly causing a fluid leak.

Inspect serpentine belts for missing ribs, wear on outside edges, cracks closer than 2 inches apart, and excessive glazing. Any of these require immediate replacement.

Serpentine or V-ribbed belts must also be properly tensioned, but these belts are usually fitted with automatic tensioners. The tensioner automatically adjusts belt tension as the V-ribbed belt stretches. These tensioners often have built-in wear indicator scales. As long as an arrow on the tensioner is located between two lines, the belt is not excessively stretched. When the arrow moves outside the lines, the belt must be replaced.

Many vehicles are equipped with plastic tensioner and idler pulleys. It is very important to inspect these for wear. They will allow belt tension to drop or even damage the belt when they wear.

8. Inspect and replace engine cooling system and heater system hoses, pipes, and fittings.

Check all cooling system hoses for loose clamps, leaks, and damage. Look for cracks, abrasions, bulges, and swelling. Check for hard spots due to heat damage from close proximity to exhaust system components. Also look for shiny spots caused by contact with accessory mounting brackets or other components. These spots may indicate weak spots that could cause a hose to burst. Check the hoses for soft or gummy areas due to contact with engine oil, power steering fluid, or automatic transmission fluid.

Squeeze each hose along its entire length to check for hard or soft areas. Listen also for crackling or crunching noises while squeezing which would indicate that the reinforcing fabric is faulty or the inner liner has deteriorated. Lower radiator hoses often contain a steel spring to prevent the hose from collapsing, so you may not be able to perform the squeeze test.

When in doubt about a hose's condition, remove it and inspect the inner liner. If the liner is cracked or otherwise deteriorated, replace the hose.

Be careful when removing a faulty hose. Aggressive twisting and pulling can damage a heater core or radiator tank. If the hose is stuck to the fitting, slit the end of the hose to make removal easier.

When installing a new hose, make sure that it fits properly. Avoid twisting or stretching the hose. A hose that is too short may fail when the engine shifts during acceleration.

9. Inspect, test, and replace thermostat, coolant by-pass, and thermostat housing.

The thermostat may be tested after it has been removed from the engine. Submerge the thermostat in a pan of water and put a thermometer in the water. Suspend the thermostat and the thermometer above the bottom of the pan. Allowing them to lie on the bottom of the pan will prevent an accurate test. Heat the water while observing the thermostat valve and the thermometer. The thermostat valve should begin to open when the temperature on the thermometer is equal to the rated temperature stamped on the thermostat. Replace the thermostat if it does not open at the rated temperature.

Always replace a thermostat with one having the correct temperature rating. Do not install a "hotter" thermostat in an attempt to speed up engine warm-up time. The engine will warm up at the same rate, but operate at a higher temperature. Do not remove a functioning thermostat from an engine that is overheating. While the engine may stop overheating, coolant will flow through the engine too quickly to absorb heat adequately. Hot spots will develop in the cooling system, especially in the cylinder heads. A cracked head can result.

Be sure to install a thermostat in the correct direction and orientation. Many thermostats have an arrow indicating which way coolant should flow through the thermostat. Some thermostats have a vent hole, "jiggle" pin, or check ball assembly mounted toward the edge of the mounting flange. This device, which allows trapped air to pass through the thermostat, must be oriented properly. In most cases, the device must point upward. Check the engine manufacturer's service manual for instructions.

Inspect the thermostat housing and by-pass hose (if equipped) for cracks, deterioration, and restrictions. Thermostat housings are often made of sheet metal or a light alloy that corrodes rapidly when coolant is not changed at the recommended intervals. Replace a deteriorated thermostat housing or by-pass hose.

10. Inspect and test coolant; drain, flush, and refill cooling system with recommended coolant; bleed air as required.

Cooling system service is a topic approached differently by technicians in different repair venues. We will offer information that should be generic to all repair technicians. During an ASE test writing workshop, the participants must all agree on the content of the test so manufacturer specific items will not appear in the test unless they are considered industry standard. It is very important to keep this in mind when taking the test.

When testing coolant, there are many methods to arrive at results but the ultimate results are the same. We want to know the protection levels of the coolant for freezing, boiling, PH, corrosion protection, and in some vehicles, nitrites. Let's take a quick look at each area.

Freezing and boiling protection are linked for the most part. Most manufacturers agree that a mixture of 50 per cent water to 50 per cent coolant provides the best of both worlds in this area and the best component protection. All manufacturers will also agree that you should be sure to use the correct coolant in the vehicle without mixing coolant types or changing to one not designed for the vehicle.

PH is a measurement of the acidity or alkaline qualities of the coolant. As coolant becomes older, it drops toward the acid end of the PH scale. Most Asian manufactured vehicles aim for around 7–9 and most American and European manufactured vehicles aim for 8–9.5 on the scale. Low PH readings can be due to a deteriorated antifreeze condition or a water-heavy blend, as water is more toward the acidic side than a coolant mix is. Very high numbers can be caused by over adding antifreeze or corrosion packages during service.

In normal use, vehicles with high output ignition systems, particularly DIS systems, will cause the coolant to become electrically charged which promotes debris in the system to adhere to metal parts and can cause radiator restriction. This can really only be corrected by replacing the antifreeze or reversing the charge in the system by using some ionizing coolant recovery systems.

Corrosion protection is added when servicing the cooling system with recovery/recycling equipment and is in the antifreeze to begin with. This is a difficult area to test and a bone of contention with manufacturers who do not support coolant recycling.

The last area that will become more critical as more diesel vehicles enter the consumer market is nitrites. When out of balance they cause small bubbles to collect on castings while the engine is running. These bubbles act like little cutters over time and carve into the casting. The vibration inherent in the diesel combustion process has been known to cause bubbles in the system to create leaks in cylinder walls. There are test strips available that detect the level of nitrites and currently only a couple of manufacturers have any specifications for them.

On some vehicles, air pockets tend to develop in the cooling system as it is filled. If these air pockets are not bled off, engine overheating and even a cracked cylinder head can result. Some engines have a bleed fitting installed on the thermostat housing or an engine coolant passage to release trapped air. Loosen this fitting until all air is removed. On engines that do not have a bleed fitting, locate the highest point in the cooling system. If this point is a hose connection, loosen the hose to bleed off air. If a high point to bleed the system is not available, there are tools available that help to push the coolant from the bottom up, displacing any air locks before the vehicle is started.

▇ 11. Inspect and replace water pump.

Check the water pump for leaking hose connections, mounting gaskets, and seals. Slow or hard-to-find leaks may be easier to find if a cooling system pressure tester is connected to the radiator filler neck.

Locate and examine the vent, or weep hole, in the water pump housing. The hole is usually in the underside of the housing, so use a small inspection mirror if necessary. If the water pump seal is leaking, coolant will usually drip from the weep hole. A very slow leak may leave only coolant residue around the hole. Replace the pump if there is evidence of coolant at the weep hole.

A defective water pump bearing may cause a growling noise at idle speed. In some cases, the bearing starts to fail after being contaminated by coolant leaking past the pump seal.

With the engine shut off, grasp the fan blades or the water pump pulley, and try to move it from side to side. This will reveal any looseness in the water pump bearing. If there is any side-to-side movement in the bearing, the water pump should be replaced.

When replacing a pump, always compare the new pump to the old one. Two pumps may look very similar, but their impellers may rotate in opposite directions. In this case, the impeller blades will be shaped differently and installing the wrong pump will cause the engine to overheat.

On many engines, some of the water pump mounting bolts extend into the block water jacket. Be sure to use the specified sealant on these bolts or coolant may leak from the engine. The bolts may also seize in place, making future servicing difficult. Refer to the manufacturer's service manual for information to determine which bolts enter the water jacket.

▇ 12. Inspect and test fan (both electrical and mechanical), fan clutch, fan shroud, air dams, and cooling fan electrical circuits; repair or replace as required.

On rear-wheel drive vehicles, the engine cooling fan is usually mounted to the water pump shaft and belt driven off the crankshaft. Plain, direct-drive fan blade assemblies should be checked for loose mounting bolts, cracked blades, and loose rivets (if fan blades are riveted to a hub). A fan assembly that has any cracks should be replaced immediately.

Fans with temperature sensitive clutches should be checked for bad bearings, leaking fluid, and seized or free-wheeling clutches. With the engine off, try to spin the fan by hand. It should spin smoothly with some resistance. If the bearing feels rough, or the fan spins without resistance, replace the clutch assembly. Grasp the fan blades and try to rock the fan from side to side. Too much play indicates a bad bearing and the clutch should be replaced. Check the bimetal coil on the front of the clutch. If it is wet or covered with dirt and grime, silicone fluid is leaking out of the fluid reservoir and the clutch should be replaced.

To check for a seized fan clutch, start the engine and observe fan speed. When the engine is cold, the fan should not pull much air through the radiator, even when the engine is revved. As the engine warms up, fan speed and noise should increase noticeably. If fan noise and speed seem excessive, stop the engine and put paint marks on the fan pulley and the back of the fan clutch. Hook up an engine timing light and start the engine. When the timing light is pointed at the back of the fan clutch, the paint marks should move relative to one another. If the paint marks stay together as engine speed is varied, the clutch is seized and must be replaced.

Front-wheel drive vehicles are usually fitted with electrically powered fans. The fan operates only when necessary. Some fans have both high and low speeds; others have just one speed. Fan operation is usually triggered by coolant temperature and/or A/C system operation. In some systems, a temperature sensitive fan switch is threaded into a radiator tank or an engine part to sense coolant temperature. When coolant temperature approaches the upper limit of the normal operating range, the switch contacts close to turn on the fan. When coolant temperature drops to a preset value, the switch contacts open to turn off the fan. Some switches supply power or ground directly to the fan motor; others activate a relay which then powers the fan. In other systems, fan operation is controlled by the PCM, which obtains coolant temperature information from its engine coolant temperature sensor. When coolant temperature reaches a preset value, the computer activates the fan and continues to monitor coolant temperature. When the temperature drops to a preset value, the computer turns off the fan. Also note that some PCM controlled fans operate at variable speeds, dependent on load.

Coolant temperature switches can be normally open or normally closed, and sensor resistance specifications vary. Refer to the vehicle manufacturer's service manual to determine how the system you are working on operates.

When a vehicle is equipped with air conditioning, turning on the system usually activates the engine cooling fan automatically. In some cases, however, the fan is not turned on until refrigerant pressure reaches a preset value. Again, refer to the vehicle manufacturer's service manual for information.

Fan shrouds and air dams are an important part of a vehicle's cooling system; they should be in place and undamaged. The purpose of a fan shroud is to allow a round fan to create a low-pressure zone behind the entire radiator core. The purpose of an air dam is to create a high-pressure zone in front of the radiator. Both components encourage air flow through the radiator under different circumstances. The fan shroud supports good air flow at low speeds and the air dam supports good air flow at highway speeds. This would lead us to keep in mind that a vehicle that overheats at highway speeds may have a problem with the air dam, and a vehicle that overheats in traffic could have a problem with the fan shroud. The most common problems are that the fan shrouds and/or air dams are broken or missing.

13. Verify proper operation of engine-related warning indicators.

Operating temperature can be indicated to the driver either by a gauge or a light. The light typically indicates only an overheat condition, while the gauge indicates approximate operating temperature. The light will have a temperature-operated switch to control it. The gauge will have a thermistor controlling it. To check light operation, make sure the light comes on during bulb check, then remove the wire from the sensor, ground it, and see if the light comes on. If it does, it is operating properly. To test the gauge, remove the wire at the sending unit, and install the appropriate resistor as indicated by the service literature, between the wire and ground. Check to see that the gauge reading indicates the appropriate temperature.

E. Fuel, Electrical, Ignition, and Exhaust Systems Inspection and Service (7 Questions)

1. Inspect, clean, or replace fuel injection system components, intake manifold, and gaskets.

Fuel and air induction system components as they relate to engine repair require us to look at maintenance and inspection of items when performing major engine repair. Let's start from the top down.

On many late model vehicles, the most significant external component of these systems is the airflow sensor. Although the air passing through the sensor is filtered, they still get dirty and can cause drivability problems for new or repaired engines that could be detrimental to ring seating or engine operating temperatures. Air meters often require cleaning with a suitable air intake solvent that removes debris and dirt that can shroud and insulate the temperature sensing bulb and cause the PCM to miscalculate air flow.

The next component in line is common to throttle body and port fuel injected vehicles and it is the throttle body. Typically we are looking for collections of dirt inside the throttle body, behind the throttle plates, and in the idle air control passages. It is key that manufacturer's recommendations be followed when servicing these because the use of the wrong cleaner could destroy the whole part. It is generally safe to wipe them out with a shop towel and in most cases this will remove deposits. Carburetor cleaner and even some induction system cleaners could remove the coatings that are designed to resist dirt causing high or erratic idle speeds.

Next in the air tract is the intake manifold. When replacing or resealing intake manifolds, it is important to check them for warpage or corrosion around water passages. Dirt deposits can generally be removed in a parts washer prior to reassembly. Always be sure to remove any electrical switches before cleaning. Most intake manifold gaskets are intended to be installed without sealers and most are very much like cylinder head gaskets. Proper torque is critical to lasting gasket performance.

The fuel components that are on the engine have to be considered. The fuel rail and pressure regulator on fuel injected vehicles, the carburetor on older vehicles, the fuel injectors, and the related sensors and actuators are all included.

Carburetors, for the most part, saw their last application in 1995. Prior to 1985, most domestic and Asian manufactured vehicles were carbureted. Carburetors and the associated vacuum lines must be checked for leakage or deterioration.

The fuel rail and pressure regulator usually only require service when a failure occurs. During repair or replacement, it is wise to replace injector o-rings to avoid leaks that are time consuming after the job is completed. Pressure regulators usually have a vacuum source to the manifold either by a line or a passage in the throttle body on throttle body injection (TBI) engines. It is critical to make sure this line or passage is clear and in good condition to avoid over-fueling conditions caused by high fuel pressure.

Fuel injectors should be checked for evidence of leakage. O-rings should be replaced and attention should be given to any wiring that may have deteriorated.

The last area includes the sensors and actuators that relate to these systems. Many vehicles have air and coolant temperature sensors mounted on the intake manifold. Thermistor-type air temp sensors should be inspected and replaced if deposits have built up on them; cleaning often damages them. The throttle position sensor and manifold pressure sensors should be handled with care during service as should all electrical wiring. Idle speed control devices that can be cleaned should be serviced while the manifold is apart. Many idle speed control devices have air hoses that attach to them. These should be inspected for cracks or leaks that could allow unfiltered air to enter the air intake tract.

2. Inspect, service, or replace air filters, filter housings, and intake ductwork.

Inspect the air filter housing and any ductwork directing fresh air to the housing. Ductwork often leads from the radiator support or inner fender to the air filter housing. Make sure that this ductwork is present and undamaged.

Check the air filter housing itself. Make sure that the housing is securely mounted and hose or duct connections are tight. On carbureted and TBI engines, make sure that the gasket between the air filter housing and the carburetor, or TBI, unit is present and in good condition. Check the housing lid or cover to make sure that it fits properly and that any seals or gasket materials are in good condition. Streaks of dust or other debris around a sealing area indicate that the seal is leaking. Make sure that all clips or wing nuts securing the lid or cover are present and working properly. On engines equipped with an airflow sensor (most multi-point injection [MPI] engines) the duct from the air filter housing to the throttle body needs to be checked for cracks and leaks. Cracks can allow the engine to draw in unmetered air causing drivability problems.

Check the air filter for damage or excessive dirt. Check the filter or filter housing for an instruction label. Follow the manufacturer's recommendations, if present, especially those for filter element replacement intervals.

3. Inspect turbocharger/supercharger; determine necessary action.

The turbocharger/supercharger and all its mounting brackets, heat shields, and ducting should be checked for damage. Replace or repair damaged or missing components.

Check the air intake side of the turbocharger/supercharger system for leaks. If there is a leak in the intake system before the compressor housing, dirt may enter the turbocharger and damage either the compressor or turbine wheel blades. When a leak is present in the intake system between the compressor wheel housing and the cylinders, turbocharger pressure is reduced.

Turbocharger/supercharger boost pressure may be tested with a pressure gauge connected to the intake manifold. Boost pressure should be tested during hard acceleration while driving the vehicle. Excessive boost pressure may be caused by a wastegate that is stuck closed, a leaking wastegate diaphragm, or a disconnected wastegate linkage. Reduced turbocharger boost pressure may be caused by a wastegate that is stuck open.

When diagnosing the cause of blue exhaust smoke on a turbocharged vehicle, first perform oil consumption diagnosis as though the engine was not turbocharged. While turbos are commonly blamed for excessive oil consumption problems, about one-half of the turbos returned under warranty are not defective. Refer to Task A.5 for more information about excessive oil consumption on turbocharged engines.

4. Test engine cranking system; determine needed repairs.

To determine a battery's state of charge, measure open circuit voltage. First remove the battery surface charge by connecting a 50-ampere load across the battery terminals for 10 seconds. Then wait 10 minutes for the battery to stabilize. Disconnect both battery cables from the battery and use a voltmeter to measure voltage across the terminals. Open circuit voltage will be at least 12.6 volts on a fully charged battery after the surface charge has been removed. If open circuit voltage is below 12.4 volts, the battery must be charged before further testing.

The best way to determine a battery's ability to deliver power is to perform a load or capacity test. During a load test, a load tester is used to discharge the battery at one-half of its stated cold cranking ampere (CCA) rating for 15 seconds. Battery voltage is recorded at the end of this time, while the load is still applied. If the test is performed with the battery at about 70°F (21°C), a battery voltage reading of 9.6V or higher indicates that the battery is in good condition.

A discharged battery can be either fast charged or slow charged. Slow charging is always preferable if time is available to do this. Slow charging allows the chemical changes that take place during charging to occur throughout the entire thickness of the battery plates instead of on the surface of the plates only. Slow charging also lessens the chances that the battery will be overheated (and permanently damaged) during charging.

The next item we want to test is the starter. We are looking for current demand and for any unusual sounds. Depending on the engine and starter design, most starters will require 150-250 amps to turn over the engine. If the starter current draw is high, there may be internal problems with it such as a dragging armature or worn bushings. External problems such as tight or seized engine-driven accessories, or even a front pump in the automatic transmission, can be a cause of higher than normal starter current draw. Listen for noise or very loud cranking noises. Some vehicles require shims to properly place the starter. If the vehicle had shims in it be sure to replace them during reassembly. If the starter current draw is low and the engine spins quickly but fails to start, suspect a jumped or broken timing chain or belt.

Battery cables are required to carry high amperage. The most effective method of testing a battery cable is with a voltage drop test. The voltmeter is placed across the cable and the engine is cranked. The meter is read while the engine is cranking. A typical voltage drop specification for the battery cable is 0.3 volts. The test should be performed on both sides of the circuit. A reading higher than specification would indicate higher than acceptable resistance. If the voltage drop is occurring at the end of the cable, the terminal may be replaced. If this method is chosen, care should be taken to ensure that the cable shortened is enough to remove all corrosion. Many technicians will replace the entire cable.

5. Inspect and replace crankcase ventilation system components.

The crankcase ventilation system serves two purposes. First, it removes crankcase blow-by and pressure. Second, it introduces the crankcase gases into the engine to be burned during the combustion process.

Inspect the system by looking for hoses that are cracked, swelled, or kinked. The side of the system with the crankcase ventilation valve in it is under vacuum; the side that connects to the air cleaner or air intake tube is the vent side. Air is drawn from the filtered intake side to ventilate the crankcase and then drawn through the valve and into the intake stream. If the ventilation valve hose is kinked, or clogged, the valve cannot open. This will result in excessive crankcase pressure that may cause gaskets and seals to leak. This may also cause an accumulation of oil in the air cleaner assembly.

6. Inspect and install ignition system components; adjust timing.

On engines fitted with a distributor, check for a cracked, worn out, or damaged distributor cap. Pull each spark plug wire from the cap one at a time and check for burned or corroded terminals. Check the spark plug cables for burned, pinched, cut, or oil-soaked insulation. Replace damaged cables. Remove the cap and check inside it. If the cap has excessively worn or corroded terminals, replace it. Check for carbon tracking and, if found, replace the cap. Check the high tension cable leading to the ignition coil. Ignition coils sometimes leak oil, which will soften and damage the cable. If oil is found, replace the ignition coil and the cable. Check the distributor rotor for a burned, pitted, or excessively worn contact.

On engines equipped with solid state ignition, check the centrifugal and vacuum advance mechanisms, if equipped. Inspect the reluctor, or pole piece, to make sure that it is not contacting the magnetic pickup or pickup coil. Replace damaged parts. The mechanical advance mechanism advances spark timing as engine speed increases. Check that the advance mechanism is not seized by grasping the rotor and attempting to turn it in the direction of rotor rotation. The rotor should move in the direction of rotation against spring pressure, but not the opposite direction. When the rotor moves, pivoted weights under the rotor or breaker plate should move outward. The vacuum advance unit controls spark advance in relation to engine load. To test the advance unit, connect a hand operated vacuum pump to the hose nipple on the unit and watch the breaker plate while operating the pump. The breaker plate should rotate in the direction opposite that of the distributor rotor as the pump is operated. If the unit will not hold vacuum, the diaphragm is damaged and the vacuum advance unit must be replaced. If the unit holds vacuum, but the breaker plate does not move, the plate may be seized. Check for a rusted pivot point or a foreign object, such as a dropped point retaining screw, that may be jammed between the plate and the distributor housing.

Remove and inspect the spark plugs. Use the appropriate socket to prevent spark plug damage. Remove the plugs when the engine is cold, especially if the engine has aluminum cylinder heads. Removing plugs from an aluminum head when the engine is hot can damage the aluminum threads. If the plugs are in good condition, apply a small amount of anti-seize compound to the spark plug threads. Thread the plugs into the head by hand to avoid cross-threading. Torque the plugs to specifications.

On most, but not all, engines fitted with a distributor type ignition system, the distributor can be rotated to adjust spark timing. Vehicles produced since 1972 have an underhood emissions label that outlines the steps necessary to adjust spark timing. Follow these instructions. In a typical sequence, the engine is brought to normal operating temperature. The vacuum advance hose (if equipped) is then disconnected and the hose is plugged. A stroboscopic timing light is connected to the #1 cylinder spark plug cable and the engine is started. Idle speed is adjusted to specifications, and the timing light is then pointed at a metal tab attached to the timing cover. If necessary, the distributor is rotated to cause a notch in the crankshaft pulley, harmonic balancer, or flywheel to align with a timing mark. On DIS inspect coil packs, plug wires, and modules for signs of arcing; replace any damaged parts. Inspect condition of crank and cam sensors. Cracked or severely oil-soaked components should be replaced.

7. Inspect and diagnose exhaust system; determine needed repairs.

Exhaust manifolds can be made of cast iron or sheet metal. Sheet metal manifolds are usually made of stainless steel. Inspect exhaust manifolds for cracks and leaks. On vehicles with computer-controlled fuel delivery systems, air entering the exhaust system through a crack or leak ahead of the oxygen sensor can cause drivability and emission control system problems. Replace a cracked manifold.

The exhaust manifold on carbureted and TBI engines may be equipped with a manifold heat control valve. This valve is closed when the engine is cold to direct hot exhaust gases to the underside of the intake manifold, directly under the carburetor or TBI unit. The gases heat the manifold, improving fuel vaporization in the cold engine. The valve opens as the engine warms up and the added heat is not needed. If the manifold heat control valve is stuck open or fails to close when the engine is cold, the engine may stumble during acceleration. If the valve is stuck in the closed position, engine power will be reduced and the intake manifold will overheat. The floor of the intake manifold may crack.

With the engine cold and shut off, check to see if the valve moves freely. Older engines use a bimetal thermostatic spring to operate the valve. Grasp the valve counterweight and rotate it back and forth. In some applications, the valve is opened and closed using a vacuum actuator. Connect a vacuum pump to the actuator and apply vacuum to test the valve. The valve shaft and bushings should be lubricated periodically with a special solvent that contains graphite. Check the vehicle manufacturer's service manual for lubricant information.

Sample Preparation Exams

INTRODUCTION

Included in this section is a series of six individual preparation exams that you can use to help determine your overall readiness to successfully pass the Engine Repair (A1) ASE certification exam. Located in Section 7 of this book you will find blank answer sheet forms you can use to designate your answers to each of the preparation exams. Using these blank forms will allow you to attempt each of the six individual exams multiple times without risk of viewing your prior responses.

Upon completion of each preparation exam, you can determine your exam score using the answer keys and explanations located in Section 6 of this book. Included in the explanation for each question is the specific task area being assessed by that individual question. This additional reference information may prove useful if you need to refer back to the task list located in Section 4 for additional support.

PREPARATION EXAM 1

1. Technician A says that it is important to test drive the customer's vehicle to verify the concern as a part of the diagnosis procedure. Technician B says it is always a good idea to check for associated technical service bulletins. Who is correct?

   A. A only
   B. B only
   C. Both A and B
   D. Neither A nor B

2. Technician A says main bearing oil clearance can be checked with Plastigauge®. Technician B says main bearing oil clearance can be checked with a feeler gauge. Who is correct?

   A. A only
   B. B only
   C. Both A and B
   D. Neither A nor B

3. Technician A says that when the fuel lines are reconnected on a replaced engine, the o-ring seals on the fuel lines must be replaced. Technician B says the injector o-rings should be replaced when they are reinstalled. Who is correct?

   A. A only
   B. B only
   C. Both A and B
   D. Neither A nor B

4. The oil filter on a customer's car is bulged out. Technician A says this could be caused by improper filter installation. Technician B says this could be caused by a stuck closed pressure regulator. Who is correct?

   A. A only

   B. B only

   C. Both A and B

   D. Neither A nor B

5. A cylinder head for an OHC engine is being inspected. The deck warpage is greater than the maximum allowed. Technician A says the head must be replaced. Technician B says the head may be able to be straightened, trued, and reused. Who is correct?

   A. A only

   B. B only

   C. Both A and B

   D. Neither A nor B

6. Technician A says that a turbocharger's boost is controlled by a wastegate. Technician B says the wastegate is controlled by engine vacuum. Who is correct?

   A. A only

   B. B only

   C. Both A and B

   D. Neither A nor B

7. All of the following are causes of low engine oil pressure EXCEPT:

   A. Worn camshaft bearings.

   B. Partially plugged oil pickup screen.

   C. Worn crankshaft bearings.

   D. Restricted pushrod passages.

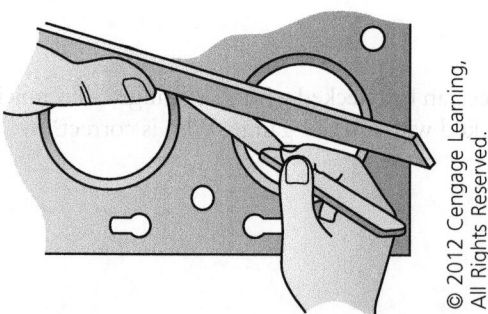

© 2012 Cengage Learning, All Rights Reserved.

8. In the figure above, what is the technician checking?

   A. Head warpage

   B. Piston protrusion

   C. Surface finish

   D. Head bolt hole alignment

9. The customer is concerned about poor performance at highway speeds. The technician talks to the owner about the car's repair history and is told that the timing belt, water pump, and crank seal have recently been replaced and the problem started afterward. What is the most likely cause for poor performance?

   A. The serpentine belt is loose.

   B. The sparkplug wires were misrouted.

   C. The camshaft and crankshaft are out of sync.

   D. The catalytic converter is restricted.

10. Technician A says the air intake hose assembly may cause poor performance if it is not properly attached. Technician B says all vehicles use a map sensor to tell the electronic control module (ECM) the amount of ingested air. Who is correct?

    A. A only

    B. B only

    C. Both A and B

    D. Neither A nor B

11. The customer's complaint is that there is blue smoke emitting from the tailpipe on initial startup in the morning, which seems to dissipate quickly. Technician A says that the blue smoke is caused by oil burning in the combustion chamber. Technician B says the oil probably is entering the combustion chamber through deteriorated valve stem seals. Who is correct?

    A. A only

    B. B only

    C. Both A and B

    D. Neither A nor B

12. All lifters in an overhead valve engine are cupped (concave). Technician A says the lifters can be replaced with no further parts replacement. Technician B says that they can be replaced with roller lifters for longer life. Who is correct?

    A. A only

    B. B only

    C. Both A and B

    D. Neither A nor B

13. Technician A says a PCV valve hose that is restricted could cause oil accumulation in the air cleaner. Technician B says a PCV valve hose that is restricted will cause a stumble on acceleration. Who is correct?

    A. A only

    B. B only

    C. Both A and B

    D. Neither A nor B

14. Technician A says that during engine removal, it is okay to open the hoses on the A/C compressor to relieve pressure so you can remove the compressor. Technician B says that the refrigerant in an A/C system must be recovered into an A/C recovery machine. Who is correct?

    A.   A only
    B.   B only
    C.   Both A and B
    D.   Neither A nor B

15. A cylinder head has been removed and the technician is inspecting the head gasket. He finds the fire ring on cylinder #2 has broken. Technician A says this could be caused by detonation in the cylinder. Technician B says this could be caused by the wrong oil being used. Who is correct?

    A.   A only
    B.   B only
    C.   Both A and B
    D.   Neither A nor B

16. The customer's complaint is that the engine cranks over quickly but does not start. Technician A says it could be caused by the battery. Technician B says it could be caused by a broken timing belt. Who is correct?

    A.   A only
    B.   B only
    C.   Both A and B
    D.   Neither A nor B

17. Technician A says corrosion on the battery terminals may cause slow cranking speed. Technician B says high resistance in the starter circuit may result in low cranking speed and high current draw. Who is correct?

    A.   A only
    B.   B only
    C.   Both A and B
    D.   Neither A nor B

18. Technician A says that when head gaskets are installed, it is important to read the orientation instructions printed on the head gasket. Technician B says an improperly oriented head gasket may restrict coolant flow to the heads. Who is correct?

    A.   A only
    B.   B only
    C.   Both A and B
    D.   Neither A nor B

19. Technician A says RTV can be used to seal the oil pan in place of a gasket. Technician B says RTV can be used on a head gasket. Who is correct?

    A. A only
    B. B only
    C. Both A and B
    D. Neither A nor B

20. A properly working catalytic converter converts HC, CO, and NOx into:

    A. O_3, H_2O, and NO
    B. H_2O, CO_2, and NO
    C. H_2O, NO, and N_2
    D. H_2O, CO_2, and N_2

21. The customer complains of a thummp and vibration when accelerating from a stop. Which of the following is the most likely cause for the condition?

    A. Out of balance tire
    B. Broken or weak motor mount
    C. Spark plug misfire
    D. A warped brake rotor

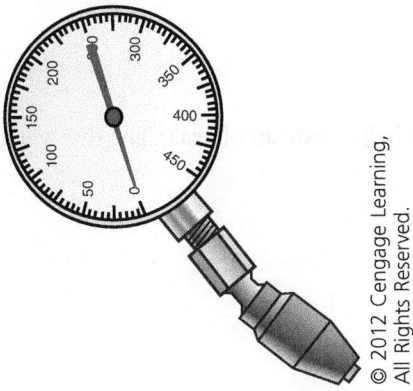

© 2012 Cengage Learning, All Rights Reserved.

22. The gauge in the figure above is used when performing a:

    A. Compression test.
    B. Cylinder power balance test.
    C. Cylinder leakage test.
    D. Vacuum test.

23. A head bolt has broken flush with the block deck surface during head removal. Technician A says the remaining bolt can be drilled and a bolt extractor can be used to remove the bolt. Technician B says the broken bolt may be completely drilled out oversize and an oversize head bolt installed. Who is correct?

    A. A only
    B. B only
    C. Both A and B
    D. Neither A nor B

24. The valve spring is being checked and fails to meet the open pressure specification. Technician A says to install a valve spring washer when reassembling the head. Technician B says to replace the spring. Who is correct?

    A. A only

    B. B only

    C. Both A and B

    D. Neither A nor B

25. Technician A says a catalytic converter may be legally removed on a vehicle that is more than 15 years old. Technician B says the catalytic converter significantly restricts exhaust flow. Who is correct?

    A. A only

    B. B only

    C. Both A and B

    D. Neither A nor B

26. Technician A says a serpentine belt that has any cracks in it must be replaced. Technician B says that serpentine belts must be retightened after break-in. Who is correct?

    A. A only

    B. B only

    C. Both A and B

    D. Neither A nor B

27. Florescent dye has been added to the crankcase to help locate an oil leak. The dye will glow when it is exposed to:

    A. A strobe light.

    B. An infrared light.

    C. A blacklight.

    D. A blue light.

28. Worn valve guides may cause all of these problems EXCEPT:

    A. Leaking of combustion gases.

    B. Excessive oil consumption.

    C. Uneven valve seating.

    D. Blowby.

29. Technician A says improper balance shaft timing can cause severe engine vibrations. Technician B says balance shafts are always timed in relation to the camshaft. Who is correct?

    A. A only

    B. B only

    C. Both A and B

    D. Neither A nor B.

30. The valve stem installed height is too tall. All of the following may be used to correct the installed height EXCEPT:
    A. Cutting the tip of the valve.
    B. Replacing the seat.
    C. Shimming the valve.
    D. Replacing the valve.

31. All of the following would be a result of a plugged catalytic converter EXCEPT:
    A. Stalling.
    B. Loss of power.
    C. Power increase.
    D. Engine overheating.

32. Technician A says an oil filter may contain an anti-drainback valve. Technician B says an oil filter will contain a bypass valve. Who is correct?
    A. A only
    B. B only
    C. Both A and B
    D. Neither A nor B

33. The customer complains of a knocking noise during cold starts in the morning that goes away within a minute of operation. Technician A says this could be caused by excessive connecting rod bearing clearance. Technician B says this could be piston slap caused by worn piston skirts. Who is correct?
    A. A only
    B. B only
    C. Both A and B
    D. Neither A nor B

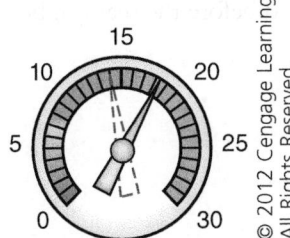

© 2012 Cengage Learning, All Rights Reserved.

34. With the engine idling, a vacuum gauge connected to the intake manifold fluctuates as shown in the figure above. These fluctuations may be caused by:
    A. Late ignition timing.
    B. A restricted exhaust system.
    C. Intake manifold leak at the throttle body.
    D. Sticky valves.

35. Technician A says that when you are measuring ring end gap, it is only necessary to measure one of each ring set to verify correct fit. Technician B says to put each individual ring in the cylinder it will be in and measure ring end gap. Who is correct?

    A. A only

    B. B only

    C. Both A and B

    D. Neither A nor B

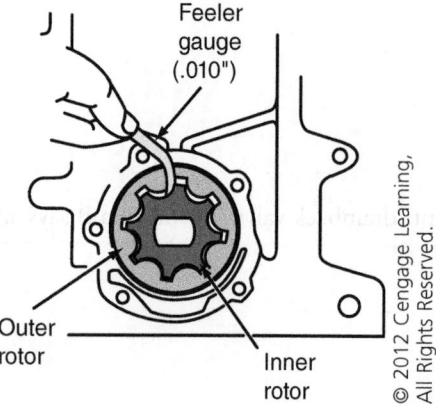

36. Technician A says that in the figure above, the oil pump drive gear-to-driven gear clearance is being checked. Technician B says the drive gear-to-pump body clearance is being checked. Who is correct?

    A. A only

    B. B only

    C. Both A and B

    D. Neither A nor B

37. Technician A says the valve seat must be refinished before any other head repairs are done. Technician B says the valve guide must be checked and repaired before the seat can be cut. Who is correct?

    A. A only

    B. B only

    C. Both A and B

    D. Neither A nor B

38. Technician A says a special puller and installer tool may be required to remove and install the harmonic balancer. Technician B says if there is deterioration of the rubber inertia ring, the balancer must be replaced. Who is correct?

    A. A only

    B. B only

    C. Both A and B

    D. Neither A nor B

39. A cylinder power balance test is being performed to determine which cylinder is causing a misfire. Technician A says when the misfiring cylinder is disabled, the RPM will drop little, or not at all. Technician B says the misfiring cylinder will cause an RPM increase when it is shorted. Who is correct?

 A. A only

 B. B only

 C. Both A and B

 D. Neither A nor B

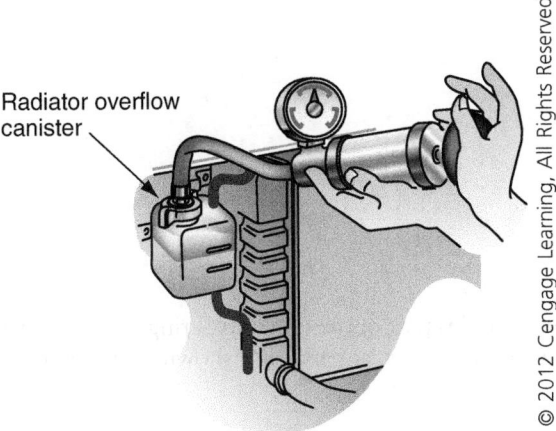

Radiator overflow canister

© 2012 Cengage Learning, All Rights Reserved.

40. The tester in the figure above may be used to test all of the following EXCEPT:

 A. Cooling system leaks.

 B. Radiator cap pressure relief valve.

 C. Heater core leaks.

 D. Coolant specific gravity.

41. Technician A says the proper place to measure piston diameter is at the crown above the first ring land. Technician B says the proper place to measure piston diameter is on the trunk, just below the piston pin. Who is correct?

 A. A only

 B. B only

 C. Both A and B

 D. Neither A nor B

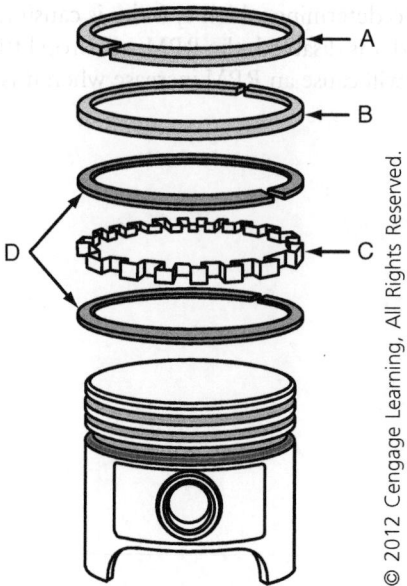

42. In the figure above, Technician A says that C is part of the oil control ring set. Technician B says that A is the top compression ring and may have a mark on it showing the top of the ring. Who is correct?

    A. A only

    B. B only

    C. Both A and B

    D. Neither A nor B

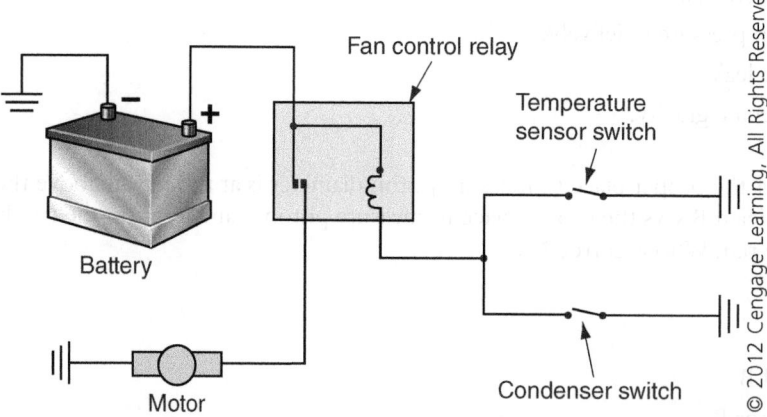

43. In the figure above, an open ground circuit on the engine temperature switch may cause:

    A. Continual cooling fan operation.

    B. The fan not to operate when the A/C is engaged.

    C. A burned out cooling fan motor.

    D. Engine overheating.

44. A cylinder leakage test has been performed and one cylinder failed with a leakage of over 50 percent. The technician notes that there are bubbles coming up in the radiator. What is the most likely source of the leakage?

    A. A leaking intake valve

    B. Worn rings

    C. Cracked head

    D. Leaking head gasket

45. A small pickup with a 4-cylinder engine is brought into the shop for a rattling noise from the front of the engine. Technician A says the noise could be caused by a timing chain that has too much slack. Technician B says the timing chain guides could be worn out. Who is correct?

    A. A only

    B. B only

    C. Both A and B

    D. Neither A nor B

46. Technician A says that when performing a cylinder leakage test, 10 percent leakage is acceptable. Technician B says the acceptable leakage is due to incomplete sealing of the rings. Who is correct?

    A. A only

    B. B only

    C. Both A and B

    D. Neither A nor B

47. Technician A says that a cylinder with 0.020 inch ring ridge can be honed and reused. Technician B says the ring ridge may have to be removed to remove the piston assembly. Who is correct?

    A. A only

    B. B only

    C. Both A and B

    D. Neither A nor B

48. An engine has a deep knock while running that increases in speed along with engine RPM. The most likely cause of the noise is:

    A. Worn lifters.

    B. A cracked flexplate.

    C. Water pump bearings.

    D. A connecting rod with too much bearing clearance.

49. Technician A says a defective radiator cap pressure valve may cause an engine to run too cold. Technician B says a defective radiator cap vacuum valve may cause the upper radiator hose to collapse. Who is correct?

    A. A only
    B. B only
    C. Both A and B
    D. Neither A nor B

50. The vehicle will not crank. Technician A says this may be caused by hydrolock in a cylinder. Technician B says the fuel pump may be defective. Who is correct?

    A. A only
    B. B only
    C. Both A and B
    D. Neither A nor B

PREPARATION EXAM 2

1. The customer's complaint is blue smoke emitting from the tailpipe on initial startup in the morning, which seems to dissipate quickly. Technician A says that the blue smoke is caused by oil burning in the combustion chamber. Technician B says the oil is probably entering the combustion chamber through deteriorated valve stem seals. Who is correct?

    A. A only
    B. B only
    C. Both A and B
    D. Neither A nor B

2. While performing a cranking compression test on a 4-cylinder engine, the technician notes that one cylinder has a pressure reading of 60 psi, 44 kPa, while the others have a reading of 135 psi, 931 kPa. Technician A says performing a cylinder leakage test will indicate where the pressure is leaking. Technician B says the vehicle has leaking valve stem seals. Who is correct?

    A. A only
    B. B only
    C. Both A and B
    D. Neither A nor B

3. The customer's complaint is excessive oil use and spots of oil on his driveway. Upon inspection, it appears that the valve cover gaskets, oil pan, and front main seal are leaking. Technician A says the problem is gasket deterioration and they must be replaced. Technician B says a defective PCV valve may be causing the leaks. Who is correct?

    A. A only
    B. B only
    C. Both A and B
    D. Neither A nor B

4. The customer says his oil pressure gauge stays low even at highway speeds. Technician A says this may be caused by worn crankshaft main bearings. Technician B says this can be caused by leaking rings. Who is correct?

    A. A only
    B. B only
    C. Both A and B
    D. Neither A nor B

5. The customer's complaint is that the engine cranks over slowly but does not start. Technician A says it could be caused by the battery. Technician B says it could be caused by a broken timing belt. Who is correct?

    A. A only
    B. B only
    C. Both A and B
    D. Neither A nor B

6. Technician A says RTV can be used to seal lower intake plenum in place of a gasket. Technician B says RTV can be used to replace a water pump gasket. Who is correct?

    A. A only
    B. B only
    C. Both A and B
    D. Neither A nor B

7. The customer complains of a vibration when accelerating that stops when reaching a steady speed. Which of the following is the most likely cause for the condition?

    A. Out of balance tire
    B. Broken or weak motor mount
    C. Bent or broken engine cooling fan
    D. A warped brake rotor

8. When preparing an engine for removal, which of the following would be done?

    A. Drain engine coolant.
    B. Drain engine oil.
    C. Disconnect fuel lines.
    D. All of the above

9. A technician is performing a cylinder leakage test and cylinder #3 has 45 percent leakage. Which is the LEAST LIKELY place the technician would look for escaping air?

    A. The cooling system
    B. The intake system
    C. The exhaust system
    D. The transmission dipstick

10. Technician A says oil accumulation in the air cleaner could be caused by worn rings. Technician B says oil accumulation in the air cleaner could be caused by a defective PCV valve or hose. Who is correct?

    A. A only
    B. B only
    C. Both A and B
    D. Neither A nor B

11. A cylinder power balance test is being performed. Technician A says a compression gauge is used for a power balance test. Technician B says a misfiring cylinder will cause a small RPM decrease or no change at all when it is shorted. Who is correct?

    A. A only
    B. B only
    C. Both A and B
    D. Neither A nor B

12. Technician A says a rod knock will become quieter as its cylinder is grounded. Technician B says a loose piston pin will cause a double click noise. Who is correct?

    A. A only
    B. B only
    C. Both A and B
    D. Neither A nor B

13. A customer says his car will not go over 50 mph. The technician installs a vacuum gauge on the engine and notes a normal, steady vacuum at idle. When he brings the RPM to 2,000 rpm and holds it there, he notes a steady drop in vacuum. Which of the following is the most likely cause?

    A. A leaking intake manifold
    B. A misfiring spark plug
    C. A plugged exhaust system
    D. Weak rings

14. There is coolant leaking from an engine compartment but the technician cannot tell where it is coming from. Technician A says dye could be put into the cooling system and the leak will show up under inspection with a blacklight. Technician B says to pressurize the cooling system and then look for leaks. Who is correct?

    A. A only
    B. B only
    C. Both A and B
    D. Neither A nor B

15. Technician A says main bearing oil clearance can be checked with Plastigauge®. Technician B says main bearing oil clearance can be checked with a feeler gage. Who is correct?

    A. A only
    B. B only
    C. Both A and B
    D. Neither A nor B

16. The spark plugs are being replaced. Technician A says platinum spark plug gap does not have to be checked. Technician B says a replacement spark plug must be of the same heat range and style as the original. Who is correct?

    A. A only
    B. B only
    C. Both A and B
    D. Neither A nor B

17. Florescent dye has been added to the crankcase to help locate an oil leak. Technician A says the dye will be visible with an infrared light at the point of the leak. Technician B says a blacklight will cause the dye to be visible. Who is correct?

    A. A only
    B. B only
    C. Both A and B
    D. Neither A nor B

18. Technician A says a light coat of RTV should be applied to a rubber valve cover gasket during installation to ensure a good seal. Technician B says RTV is a gasket maker and should not be applied to the rubber valve cover gasket. Who is correct?

    A. A only
    B. B only
    C. Both A and B
    D. Neither A nor B

19. The threads in a camshaft tower hole have been damaged. Technician A says the threads may be tapped with the proper tap size and thread pitch to restore the threads. Technician B says the threads may have to be restored using a HeliCoil thread insert. Who is correct?

    A. A only
    B. B only
    C. Both A and B
    D. Neither A nor B

20. An oil pan is being reinstalled on the engine. Technician A says that the oil pan can be sealed using RTV. Technician B says that a new gasket must be installed. Who is correct?

    A. A only
    B. B only
    C. Both A and B
    D. Neither A nor B

21. Technician A says balance shafts rotate in the opposite direction of the crankshaft. Technician B says the balance shafts are always timed in relation to the crankshaft. Who is correct?

    A. A only
    B. B only
    C. Both A and B
    D. Neither A nor B

22. The customer is concerned about poor performance at highway speeds. The technician talks to the owner about the car's repair history and is told that the timing belt, water pump, and crank seal have recently been replaced, and the problem started afterward. What is the most likely cause for poor performance?

    A. The serpentine belt is loose.

    B. The sparkplug wires were misrouted.

    C. The camshaft and crankshaft are out of sync.

    D. The catalytic converter is restricted.

23. During disassembly of the block, the technician notes that main bearing wear is greater on the first journal's top bearing and the bottom bearing of the last journal. Technician A says this may have been caused by accessory belts being overtightened. Technician B says this was caused by oil starvation. Who is correct?

    A. A only

    B. B only

    C. Both A and B

    D. Neither A nor B

24. Technician A says that when the fuel lines are reconnected on a replaced engine, the o-ring seals on the fuel lines must be replaced. Technician B says the injector o-rings should be replaced when they are reinstalled. Who is correct?

    A. A only

    B. B only

    C. Both A and B

    D. Neither A nor B

25. All of the following could prevent a starter solenoid from engaging EXCEPT:

    A. A bent armature shaft.

    B. A shorted clutch safety switch.

    C. An open ignition switch.

    D. An open in the neutral safety switch.

26. Technician A says during disassembly of the engine block, all parts must be kept in order for later inspection. Technician B says connecting rod caps should be marked to identify which cylinder they came out of. Who is correct?

    A. A only

    B. B only

    C. Both A and B

    D. Neither A nor B

27. Technician A says a vacuum gauge reading of 16 to 21 in. Hg at idle is normal. Technician B says a vacuum gauge reading that is steady but low could indicate retarded valve timing. Who is correct?

    A. A only

    B. B only

    C. Both A and B

    D. Neither A nor B

28. Technician A says too tall of an installed valve stem height could cause burned valves.
    Technician B says too short of an installed valve stem height could cause poor performance.
    Who is correct?

    A. A only
    B. B only
    C. Both A and B
    D. Neither A nor B

29. Technician A says that cylinder walls that do not require boring can be cleaned and reused
    without further servicing. Technician B says the cylinder walls should be deglazed and
    cleaned to provide oil retention during break-in. Who is correct?

    A. A only
    B. B only
    C. Both A and B
    D. Neither A nor B

30. Technician A says talking to the customer to make sure you understand his complaint is a
    good practice. Technician B says it is important to test drive the vehicle to duplicate the
    customer's complaint. Who is correct?

    A. A only
    B. B only
    C. Both A and B
    D. Neither A nor B

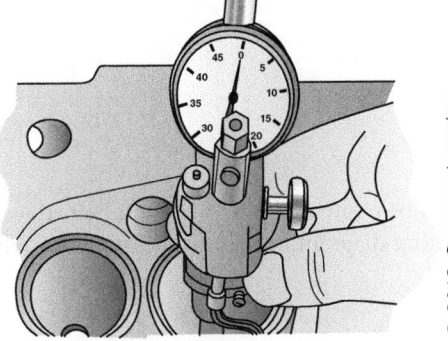

© 2012 Cengage Learning,
All Rights Reserved.

31. In the figure above, the technician is most likely checking:

    A. Valve guide wear.
    B. Valve stem installed height.
    C. Valve seat concentricity.
    D. Valve seat angle.

32. An upper radiator hose that collapses as the engine cools down can be caused by:

 A. A defective radiator cap pressure seal.

 B. A faulty upper radiator hose.

 C. A stuck open thermostat.

 D. A defective radiator cap vacuum valve.

33. Technician A says during teardown, the main and rod bearings should be kept in order to identify abnormal wear. Technician B says the main bearing bore should be checked for misalignment. Who is correct?

 A. A only

 B. B only

 C. Both A and B

 D. Neither A nor B

34. An excessive sulfur smell in the exhaust of a vehicle with a catalytic converter can be an indication of:

 A. A lean fuel mixture.

 B. Coolant leaking into the combustion chamber.

 C. A rich fuel mixture.

 D. A vacuum leak.

35. Technician A says that when heads are removed from the block, it does not matter how the head bolts are removed. Technician B says it is best to remove the heads while they are still warm to prevent warping on the deck surface. Who is correct?

 A. A only

 B. B only

 C. Both A and B

 D. Neither A nor B

36. Technician A says that some serpentine belts can be easily misrouted causing belt squeal. Technician B says most vehicles have a belt routing diagram in the owner's manual. Who is correct?

 A. A only

 B. B only

 C. Both A and B

 D. Neither A nor B

37. When doing a compression test, the results for all cylinders are all even, but lower than the specified compression pressure. This could indicate:

 A. A blown head gasket.

 B. Worn rings and cylinders.

 C. A cracked head.

 D. Carbon buildup on the pistons.

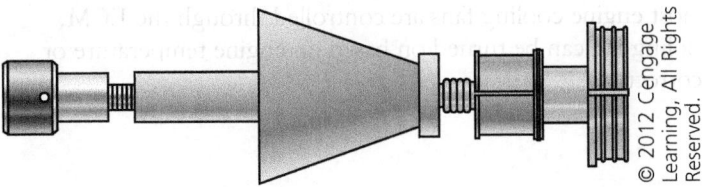

© 2012 Cengage Learning, All Rights Reserved.

38. The tool shown in the figure above is used to:

    A. Remove crankshaft bearings.

    B. Remove camshaft bearings only.

    C. Remove pistons.

    D. Remove and install camshaft bearings.

39. A vehicle is brought in with excessive oil consumption. Technician A says lack of regular oil changes can cause the oil rings to stick. Technician B says a burned valve can cause excessive oil consumption. Who is correct?

    A. A only

    B. B only

    C. Both A and B

    D. Neither A nor B

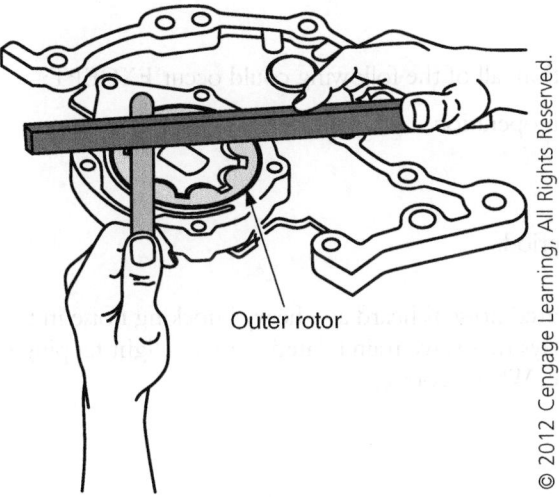

Outer rotor

© 2012 Cengage Learning, All Rights Reserved.

40. The technician in the figure above is most likely:

    A. Checking oil pump drive gear-to-driven gear clearance.

    B. Checking oil pump housing trueness.

    C. Checking oil pump gear-to-body clearance.

    D. Checking oil pump pressure relief valve.

41. Technician A says that most engine cooling fans are controlled through the ECM. Technician B says the cooling fan can be turned on based on engine temperature or A/C selection. Who is correct?

    A. A only

    B. B only

    C. Both A and B

    D. Neither A nor B

42. The technician is preparing to check an aluminum head for cracks. All of the following are acceptable methods EXCEPT:

    A. Dye.

    B. Pressure testing in a water tank.

    C. Visual inspection.

    D. Magnafluxing.

43. Which of these would be the most likely tool used to measure ring end gap?

    A. Outside micrometer

    B. Feeler gauge

    C. Dial caliper

    D. Dial indicator

44. If a thermostat fails in the open position, all of the following could occur EXCEPT:

    A. Erratic computer control system operation.

    B. Poor fuel economy.

    C. Loss of coolant.

    D. Longer than normal warmup period.

45. Technician A says that valve train related noise is heard as a heavy knocking noise in the middle of the engine. Technician B says that valve train related noise is a light tapping or rattling noise in the top of the engine. Who is correct?

    A. A only

    B. B only

    C. Both A and B

    D. Neither A nor B

46. Technician A says balance shafts may be mounted above the camshaft. Technician B says the balance shafts may be mounted on the bottom of the engine. Who is correct?

    A. A only

    B. B only

    C. Both A and B

    D. Neither A nor B

47. Technician A says an ethylene glycol and water mixture raises the freezing point of the coolant. Technician B says an ethylene glycol and water mixture lowers the boiling point of the coolant. Who is correct?

    A. A only

    B. B only

    C. Both A and B

    D. Neither A nor B

48. Upon head removal, the technician notes the coolant passage holes in the head gasket have corroded out to the same size as the block and head passages. Technician A says this could cause engine overheating at high speeds. Technician B says the holes in the gasket are substantially smaller than the coolant passage holes in the block and head. Who is correct?

    A. A only

    B. B only

    C. Both A and B

    D. Neither A nor B

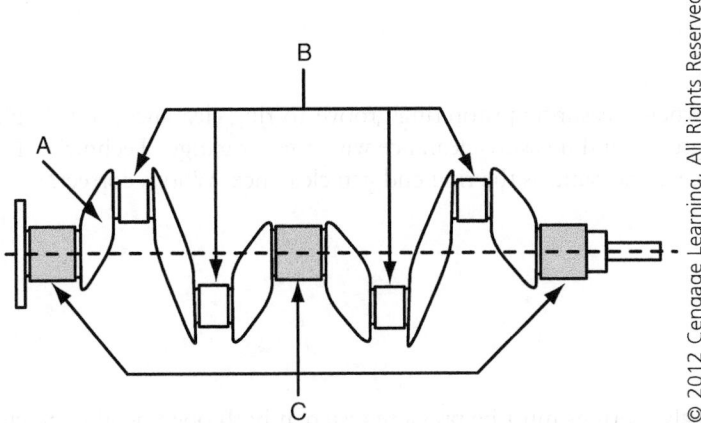

© 2012 Cengage Learning, All Rights Reserved.

49. Referring to the crankshaft in the figure above, which of the statements below is the LEAST LIKELY to be true?

    A. B indicates the rod journals.

    B. C indicates the crankshaft thrust surface.

    C. A indicates cross drilled oil supply holes.

    D. C indicates the main bearing journals.

50. Technician A says TTY bolts provide a more uniform clamping force when compared to conventional head bolts. Technician B says TTY bolts are tightened to a specified torque, then rotated a specified number of degrees. Who is correct?

    A. A only

    B. B only

    C. Both A and B

    D. Neither A nor B

PREPARATION EXAM 3

1. Technician A says a reverse flow cooling system will warm an engine up quicker than conventional flow. Technician B says a reverse flow cooling system directs coolant to the block and then to the head. Who is correct?

   A. A only

   B. B only

   C. Both A and B

   D. Neither A nor B

2. Technician A says that it is important to test drive the customer's vehicle to verify the concern as a part of the diagnosis procedure. Technician B says it is always a good idea to check for associated technical service bulletins. Who is correct?

   A. A only

   B. B only

   C. Both A and B

   D. Neither A nor B

3. Technician A says that when measuring piston ring groove-to-ring clearance, you should place the ring into the groove and measure clearance with a feeler gauge. Technician B says the ring groove clearance is the same as the ring end gap clearance. Who is correct?

   A. A only

   B. B only

   C. Both A and B

   D. Neither A nor B

4. Technician A says that valve springs must be pressure tested at both open height and closed height. Technician B says if the pressure is low at the closed height but good at the open height, the spring can be reused. Who is correct?

   A. A only

   B. B only

   C. Both A and B

   D. Neither A nor B

5. After performing a compression test on a 4-cylinder engine, the results show low pressure on cylinders #1 and #3. Technician A says this could be caused by a blown head gasket between the cylinders. Technician B says a cylinder leakage test would help pinpoint the problem. Who is correct?

   A. A only

   B. B only

   C. Both A and B

   D. Neither A nor B

6. Technician A says a power balance test is used to pinpoint the cause of low compression test results. Technician B says a power balance test is used to pinpoint an underperforming cylinder. Who is correct?

    A. A only
    B. B only
    C. Both A and B
    D. Neither A nor B

7. Technician A says an oil leak at the rear main seal could be caused by worn rings. Technician B says an oil leak at the rear main seal could be caused by a defective PCV valve or hose. Who is correct?

    A. A only
    B. B only
    C. Both A and B
    D. Neither A nor B

8. A technician is performing a cylinder leakage test and cylinders #3 and #4 have 65 percent leakage. Which is the LEAST LIKELY place the technician would look for escaping air?

    A. The adjacent cylinder
    B. The intake system
    C. The exhaust system
    D. The transmission dipstick

9. Technician A says the valve stem installed height is measured from the bottom of the valve guide to the top of the installed valve. Technician B says the valve stem installed height is measured from the spring seat to the tip of the installed valve. Who is correct?

    A. A only
    B. B only
    C. Both A and B
    D. Neither A nor B

10. Technician A says a rod bearing that is worn will sound like a deep knock at the bottom of the engine. Technician B says a loose piston pin will cause a double click noise. Who is correct?

    A. A only
    B. B only
    C. Both A and B
    D. Neither A nor B

11. If a block is to be rebuilt, all of the following parts should be removed during teardown EXCEPT:

    A. Core (freeze) plugs.
    B. Oil galley plugs.
    C. In-block camshaft bearings.
    D. Cylinder liners.

12. The exhaust coming from the tailpipe is blue in color. Technician A says this is caused by the engine running too rich. Technician B says this is caused by coolant entering the combustion chamber. Who is correct?

    A. A only

    B. B only

    C. Both A and B

    D. Neither A nor B

13. The technician installs a vacuum gauge on the engine and notes a normal reading at idle, but when the RPM is raised, the needle fluctuates rapidly between 12 and 24 in. Hg. Which of the following is the most likely cause?

    A. A leaking intake manifold

    B. A misfiring spark plug

    C. Weak valve springs

    D. Weak rings

14. Technician A says on a vehicle with a map sensor, the air intake ductwork is not necessary. Technician B says that intact air intake ductwork is critical to engine longevity. Who is correct?

    A. A only

    B. B only

    C. Both A and B

    D. Neither A nor B

15. An engine experiences very low oil pressure. Technician A says this may be caused by a stuck closed pressure relief valve. Technician B says low oil pressure will cause valve train rattle. Who is correct?

    A. A only

    B. B only

    C. Both A and B

    D. Neither A nor B

16. Technician A says that hot tanking an engine block after boring is sufficient cleaning before reassembly. Technician B says that cleaning the bores and block with hot soapy water and a bore brush ensures that all machining particles are removed. Who is correct?

    A. A only

    B. B only

    C. Both A and B

    D. Neither A nor B

17. Technician A says a running (dynamic) compression test is used to check cylinder breathing. Technician B says a running (dynamic) compression test is used to check cylinder sealing. Who is correct?

    A. A only
    B. B only
    C. Both A and B
    D. Neither A nor B

18. Technician A says that incorrect valve adjustment may cause bent valves. Technician B says incorrect valve adjustment will not affect performance. Who is correct?

    A. A only
    B. B only
    C. Both A and B
    D. Neither A nor B

19. Crankshaft inspection should include all of the following EXCEPT:

    A. Rod journal diameter and taper.
    B. Main journal diameter and taper.
    C. Dial indicator check for warpage.
    D. Crankshaft length.

20. Technician A says that a timing adjustment check is done with a timing light. Technician B says that timing adjustment may require a scan tool. Who is correct?

    A. A only
    B. B only
    C. Both A and B
    D. Neither A nor B

21. The results of a cylinder cranking compression test on a 4-cylinder engine shows low compression on cylinder #3. The technician squirts a tablespoon of oil in the cylinder and retests. The compression reading increases by 40 percent. The increase in compression indicates:

    A. A leaking intake valve.
    B. A blown head gasket.
    C. A burned exhaust valve.
    D. Worn rings.

22. A vehicle with 110,000 miles on it is brought to the shop to be checked for low oil pressure. Technician A says the oil pressure should first be checked with a mechanical gauge to eliminate the pressure sending unit as the problem. Technician B says low oil pressure can be caused by worn crankshaft main bearings. Who is correct?

    A. A only
    B. B only
    C. Both A and B
    D. Neither A nor B

23. An 8-cylinder block is being inspected and it is noted that cylinder #4 has a small crack in the middle of the bore. Technician A says that the cylinder can be sleeved to be repaired. Technician B says the crack can be welded and bored. Who is correct?

    A. A only
    B. B only
    C. Both A and B
    D. Neither A nor B

24. The exhaust outlet of a turbocharger is coated with oil. Which of the following is the most likely cause?

    A. Leaking valve stem seals
    B. A plugged PCV system
    C. Leaking turbocharger oil seals
    D. Worn engine rings

25. Timing belt replacement is being discussed. What is the first thing that should be done?

    A. Remove the harmonic balancer.
    B. Remove the water pump.
    C. Align all timing marks.
    D. Release the timing belt tensioner.

26. During a power balance test on a port fuel-injected engine, one cylinder is found to have virtually no RPM change. Which of these is the most likely cause?

    A. A faulty crankshaft position sensor
    B. A vacuum leak at the throttle body
    C. A defective plug wire
    D. A faulty camshaft position sensor

27. Technician A says that a cylinder bore that has a deep scratch can be repaired using a dingleberry hone. Technician B says that a cylinder bore that has 0.008 inch taper does not need to be bored. Who is correct?

    A. A only
    B. B only
    C. Both A and B
    D. Neither A nor B

28. Technician A says that a 13 lb. radiator cap that is being replaced could be replaced with a 15 lb. cap for better protection. Technician B says a 13 lb. cap that releases pressure at 10 lbs. when being tested is OK to be used. Who is correct?

    A. A only
    B. B only
    C. Both A and B
    D. Neither A nor B

29. Technician A says a stuck open thermostat can cause high pressure in the cooling system. Technician B says a defective radiator pressure cap can cause a burst upper radiator tank. Who is correct?

    A. A only
    B. B only
    C. Both A and B
    D. Neither A nor B

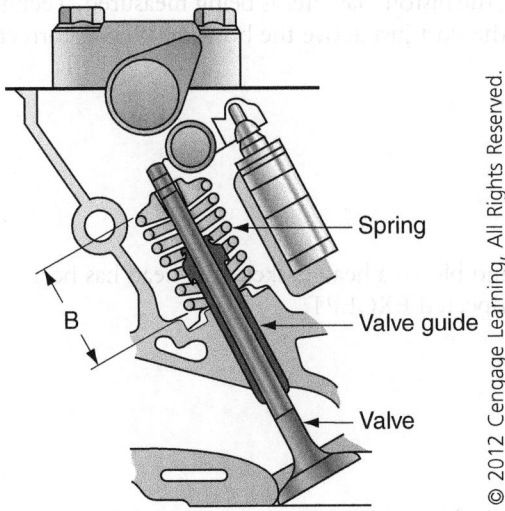

Spring

Valve guide

Valve

B

© 2012 Cengage Learning, All Rights Reserved.

30. Measurement B in the figure above is more than specified. Technician A says it can be corrected by inserting a washer below the valve spring. Technician B says this could case improper valve seating. Who is correct?

    A. A only
    B. B only
    C. Both A and B
    D. Neither A nor B

31. Technician A says a serpentine belt with three or more cracks within a 1" space should be replaced. Technician B says a smooth surface pulley that is uneven on the belt contact surface should be replaced. Who is correct?

   A. A only

   B. B only

   C. Both A and B

   D. Neither A nor B

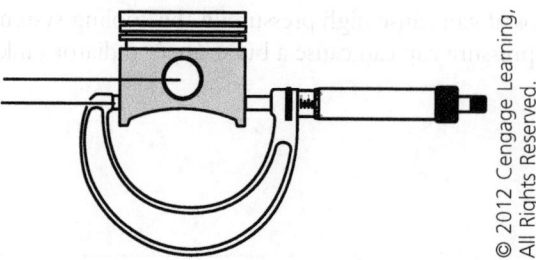

© 2012 Cengage Learning, All Rights Reserved.

32. Technician A says that in the figure above, the piston diameter is being measured. Technician B says the piston should be measured on the skirt just above the bottom. Who is correct?

   A. A only

   B. B only

   C. Both A and B

   D. Neither A nor B

33. An overhead cam engine has overheated and blown a head gasket. The head has been removed. All of the following would be inspected EXCEPT:

   A. Head deck for warpage.

   B. Camshaft bore for warpage.

   C. Camshaft followers for damage.

   D. Head for cracks.

34. Technician A says a pushrod engine's camshaft should be checked for lobe wear. Technician B says the camshaft should be checked for warping. Who is correct?

   A. A only

   B. B only

   C. Both A and B

   D. Neither A nor B

35. The technician is checking the coolant hoses on an engine. Technician A says they should be checked for cracks. Technician B says if the hose crunches when squeezed, there are deposits inside the hose and it should be replaced. Who is correct?

   A. A only

   B. B only

   C. Both A and B

   D. Neither A nor B

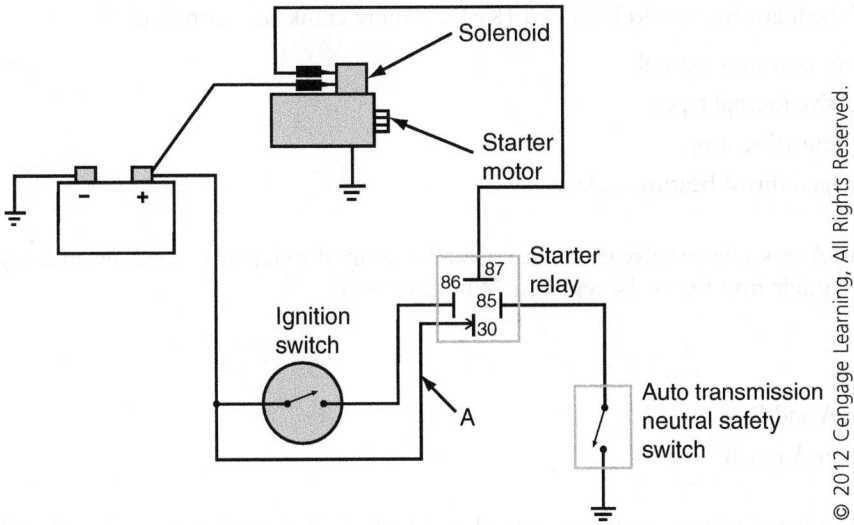

© 2012 Cengage Learning, All Rights Reserved.

36. Technician A says an open in the wiring at point A in the figure above will prevent the starter from engaging. Technician B says if the neutral safety switch is open, the starter will not engage. Who is correct?

    A. A only

    B. B only

    C. Both A and B

    D. Neither A nor B

37. When a valve spring is checked for warpage, it is:

    A. Rolled on a flat surface.

    B. Compressed to see if it bows out.

    C. Placed on a flat surface standing up against a straightedge and rotated.

    D. Measured for free length.

38. When the coolant is being tested, which of the following is the LEAST LIKELY to be tested for?

    A. Freeze protection

    B. pH level

    C. Flow volume

    D. Electrical current

39. Three types of valve stem seals are:

    A. Lip, o-ring, and floating.

    B. O-ring, positive, and box.

    C. Positive, umbrella, and lip.

    D. Positive, o-ring, and umbrella.

40. Which of the following would LEAST LIKELY require crankshaft grinding?

    A. An out of round journal
    B. Excessive journal taper
    C. Fine journal scoring
    D. Damaged thrust bearing surface

41. Technician A says a worn valve guide insert can be knurled to repair it. Technician B says a worn valve guide insert must be replaced. Who is correct?

    A. A only
    B. B only
    C. Both A and B
    D. Neither A nor B

42. Cylinder heads are being reinstalled on the short block. Technician A says the head bolt holes should be chased with a tap before installation. Technician B says it is necessary to lubricate the head bolt threads before installation. Who is correct?

    A. A only
    B. B only
    C. Both A and B
    D. Neither A nor B

43. Technician A says that after a 45 degree valve seat has been cut, a 30 degree angle is used to position the valve-to-seat contact. Technician B says a 60 degree angle is used to position the valve-to-seat contact. Who is correct?

    A. A only
    B. B only
    C. Both A and B
    D. Neither A nor B

44. An engine cranks over slowly and will not start. Technician A says the engine has to crank at 450 rpm to start. Technician B says the vehicle may have a weak battery. Who is correct?

    A. A only
    B. B only
    C. Both A and B
    D. Neither A nor B

45. There is coolant lost from an engine but the technician cannot tell where the leak is. Technician A says to pressurize the cooling system to 25 psi, then check for leaks. Technician B says small leaks may only be obvious immediately after a hot shutdown. Who is correct?

    A. A only
    B. B only
    C. Both A and B
    D. Neither A nor B

46. During a vacuum test, it is noted that the needle oscillates rapidly as the engine RPM increases but it is steady and normal at idle. What would be the most likely cause?

    A. Worn rings
    B. Sticky valves
    C. Weak valve springs
    D. A clogged exhaust system

© 2012 Cengage Learning, All Rights Reserved.

47. Technician A says the figure above shows the two halves of a one-piece main and thrust bearing. Technician B says there are two of these in an engine. Who is correct?

    A. A only
    B. B only
    C. Both A and B
    D. Neither A nor B

48. The technician is preparing to install the cylinder head on a 4-cylinder OHC engine with the cam installed. Which of these operations is the LEAST LIKELY to be performed?

    A. Run a bottoming tap through the head bolt holes.
    B. Apply a light coating of oil on the head bolts and their washers.
    C. Put #1 piston at TDC.
    D. Clean the deck surfaces of any oil residue.

49. The threads in a water pump mounting hole have been damaged. Technician A says the threads may be drilled out oversize and tapped with the proper tap size and thread pitch to accept a larger diameter bolt. Technician B says the threads may be restored using a HeliCoil thread insert. Who is correct?

    A. A only
    B. B only
    C. Both A and B
    D. Neither A nor B

50. A running (dynamic) compression test has been performed after a cranking compression test was done on a 4-cylinder engine. The results indicated cylinder #3 has substantially lower running compression than the other three. Cranking compression was within $+/- 10$ percent for all four cylinders. The running compression test results for cylinder #3 indicates:

    A. A restriction in the exhaust for that cylinder.
    B. A restriction in the intake air on that cylinder.
    C. A stopped-up catalytic converter.
    D. A plugged air filter.

PREPARATION EXAM 4

1. The customer says his oil pressure gauge reads low even at highway speeds. Technician A says this may be caused by worn crankshaft main bearings. Technician B says this can be caused by leaking rings. Who is correct?

    A. A only
    B. B only
    C. Both A and B
    D. Neither A nor B

2. The technician is diagnosing an intermittent no-crank problem. Which of the following would be the LEAST LIKELY to cause this problem?

    A. A poor connection at the battery positive post
    B. High resistance in the starter ground circuit
    C. Hydro-locked engine
    D. A worn ignition switch

3. When preparing an engine for removal, which of the following would be done?
    A. Drain engine coolant.
    B. Drain engine oil.
    C. Disconnect fuel lines.
    D. All of the above

4. The customer with a 1994 model vehicle complains of a strong fuel odor when he walks past the front of the car. Which of the following would be the LEAST LIKELY cause?
    A. Leaking injector seals
    B. A saturated evaporative emissions charcoal canister
    C. Cracked fuel line o-rings
    D. A leaking fuel pump

5. Technician A says extended life coolants provide rust and freeze protection for 5 years/ 150,000 miles. Technician B says extended life coolants provide rust and corrosion protection for 5 years/150,000 miles. Who is correct?
    A. A only
    B. B only
    C. Both A and B
    D. Neither A nor B

6. The technician is performing a cylinder cranking compression test. Technician A says a variance of 35 percent is acceptable. Technician B says there should be no more than 20 percent variance between the highest and lowest cylinder results. Who is correct?
    A. A only
    B. B only
    C. Both A and B
    D. Neither A nor B

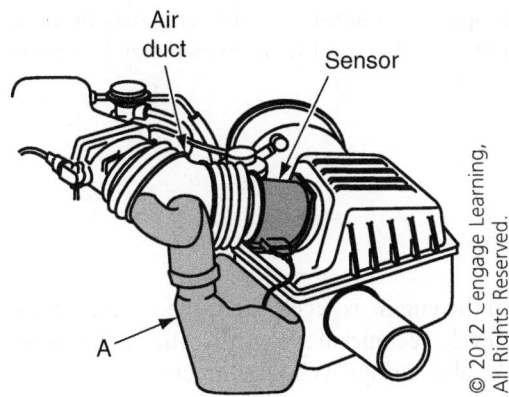

7. In the figure above, Technician A says A is a water separator. Technician B says it is a resonator to reduce intake moan. Who is correct?

    A. A only

    B. B only

    C. Both A and B

    D. Neither A nor B

8. After doing a compression test, it is determined that an engine has a blown head gasket. Which of the following is the LEAST LIKELY result?

    A. Higher than normal compression readings

    B. Oil that is a milky brown color

    C. Bubbles in the radiator

    D. White exhaust smoke

9. A sweet smell and white smoke in the exhaust of a vehicle with a catalytic converter can be an indication of:

    A. A lean fuel mixture.

    B. Coolant leaking into the combustion chamber.

    C. A rich fuel mixture.

    D. A vacuum leak.

10. Technician A says coolant with rust in it may cause the water pump impeller to fail. Technician B says the water pump impeller may be made of plastic. Who is correct?

    A. A only

    B. B only

    C. Both A and B

    D. Neither A nor B

11. The catalytic converter is overheating causing it to glow bright red and heat the passenger's floorboard. There is also a strong sulfur odor from the exhaust. The cause of this could be:

    A. Improper coolant in the engine.

    B. A vacuum leak.

    C. Too low a fuel octane.

    D. A rich fuel mixture.

12. Technician A says a lower-than-normal idle speed in a fuel-injected engine may be caused by a broken vacuum line. Technician B says an intake manifold vacuum leak may cause a misfire at idle and lower engine speeds. Who is correct?

    A. A only
    B. B only
    C. Both A and B
    D. Neither A nor B

13. During a vacuum test, it is noted that when the engine is accelerated, the vacuum drops to near zero, then climbs back to the normal level. Technician A says that the piston rings are worn. Technician B says this indicates weak valve springs. Who is correct?

    A. A only
    B. B only
    C. Both A and B
    D. Neither A nor B

14. Technician A says that some RTV sealants may be harmful to an oxygen sensor. Technician B says that RTV is an anaerobic sealer. Who is correct?

    A. A only
    B. B only
    C. Both A and B
    D. Neither A nor B

15. A front-wheel drive vehicle's engine jumps during acceleration and hits the underside of the hood. The motor mounts are inspected and although none of them appear to be broken, the front mount is coated with oil. Technician A says the motor mount was intact and was not the cause of engine movement. Technician B says the oil came from the fluid-filled hydraulic motor mount, which would allow excessive engine movement. Who is correct?

    A. A only
    B. B only
    C. Both A and B
    D. Neither A nor B

16. Reduced turbocharger boost pressure may be caused by a:

    A. Wastegate valve stuck open.
    B. Leaking wastegate diaphragm.
    C. Disconnected wastegate linkage.
    D. Wastegate stuck closed.

17. A heater hose is being replaced and the technician finds it is stuck on the hose nipple. Which of the following is the best solution to the problem?

    A. Use pliers to twist the hose off.
    B. Run a screwdriver between the hose and the nipple and pry it off.
    C. Twist the hose by hand and pull on it at the same time.
    D. Slice the hose lengthwise and peel it off.

18. The technician is testing for a restricted catalytic converter. Technician A says a vacuum gauge may be used. Technician B says a temperature sensing probe can be used. Who is correct?

    A. A only
    B. B only
    C. Both A and B
    D. Neither A nor B

19. After a vehicle sits overnight, it has a light tapping noise when it is first started that disappears after a short time. The most likely cause would be:

    A. Low oil level.
    B. Worn rod bearings.
    C. Excessive lifter leakdown.
    D. Weak oil pump.

20. Technician A says a running (dynamic) compression test should be done with the spark disabled and the fuel injector for that cylinder unplugged. Technician B says that a running compression tests results at idle should be about 50 percent of cranking compression results. Who is correct?

    A. A only
    B. B only
    C. Both A and B
    D. Neither A nor B

21. When the rod bearings on an engine were inspected, it was noted that one rod bearing set had wear on one side of the lower bearing and on the opposite side of the upper bearing. This could be caused by:

    A. Lack of lubrication.
    B. A tapered rod journal.
    C. A twisted connecting rod.
    D. A warped crankshaft.

22. A cylinder power balance test is performed on a rough running 1998 model 8-cylinder engine with four coil packs. Two cylinders on opposite heads show no RPM drop. Technician A says the ignition coil for those two cylinders may be defective. Technician B says a camshaft out of sync with the crankshaft can cause this. Who is correct?

    A. A only
    B. B only
    C. Both A and B
    D. Neither A nor B

23. An air filter that is plugged with debris could cause all of the following EXCEPT:

    A. Poor performance.
    B. Sluggish acceleration.
    C. Higher fuel economy.
    D. Lower fuel economy.

24. Valve springs are being inspected. All of the following would be checked EXCEPT:

    A. Spring free height is checked.

    B. Checked for open tension.

    C. Checked for closed tension.

    D. Spring diameter is checked.

25. The ring gear teeth of a flywheel are badly worn in multiple spots. Technician A says this could cause poor starter drive gear engagement. Technician B says this could cause a lower than normal starter current draw. Who is correct?

    A. A only

    B. B only

    C. Both A and B

    D. Neither A nor B

26. A cylinder leakage test has been performed and the results were:

Cyl. #1	Cyl. #2	Cyl. #3	Cyl. #4
15%	95%	95%	10%

    Technician A says these results indicate a blown head gasket. Technician B says Cyl. #1 and Cyl. #4 are within acceptable limits. Who is correct?

    A. A only

    B. B only

    C. Both A and B

    D. Neither A nor B

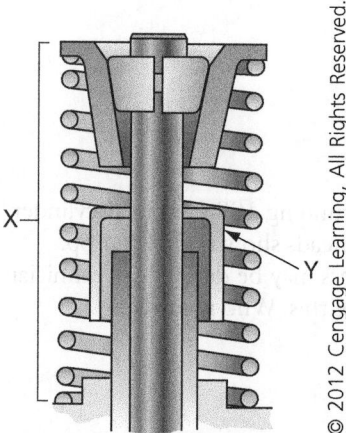

© 2012 Cengage Learning, All Rights Reserved.

27. Technician A says that X in the figure above can be replaced without removing the head. Technician B says the head must be removed to replace Y in the figure. Who is correct?

    A. A only

    B. B only

    C. Both A and B

    D. Neither A nor B

28. A vehicle is being checked for coolant loss. There is no sign of coolant loss in the engine compartment. Technician A says coolant loss could be caused by a bad radiator cap. Technician B says the heater core may be leaking. Who is correct?

    A. A only
    B. B only
    C. Both A and B
    D. Neither A nor B

29. The vehicle cranks normally but is consistently slow to start. Which of the following is the LEAST LIKELY cause?

    A. A partially plugged fuel filter
    B. A failed fuel pump check valve
    C. Leaking injectors
    D. A defective fuel pressure regulator

30. Technician A says that excessive wear on the lower halves of the main bearing inserts is caused by prolonged idle operation. Technician B says this wear can be caused by prolonged high RPM. Who is correct?

    A. A only
    B. B only
    C. Both A and B
    D. Neither A nor B

31. A cylinder cranking compression test has been done; the results were all well below manufacturer's specifications. Technician A says a wet cylinder compression test should be done to eliminate valves as the possible cause. Technician B says a power balance test should be done to identify the problem. Who is correct?

    A. A only
    B. B only
    C. Both A and B
    D. Neither A nor B

32. The head has been sent to the machine shop to be resurfaced. Technician A says that a belt sander will true the surface and smooth it. Technician B says the surface finish is critical to proper head gasket sealing and life. Who is correct?

    A. A only
    B. B only
    C. Both A and B
    D. Neither A nor B

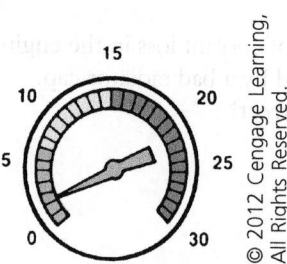

© 2012 Cengage Learning, All Rights Reserved.

33. A vacuum test has been done and the results at 2,500 rpm are shown in the figure above. Technician A says the very low readings could be caused by a sticky valve. Technician B says the low, steady readings are probably the result of a severely restricted exhaust system. Who is correct?

    A. A only

    B. B only

    C. Both A and B

    D. Neither A nor B

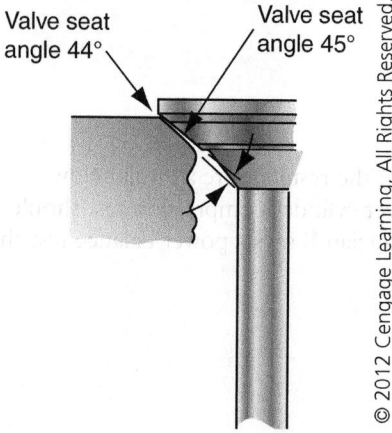

© 2012 Cengage Learning, All Rights Reserved.

34. In the figure above, Technician A says seat width is being shown. Technician B says the interference angle between the seat and valve face is being shown. Who is correct?

    A. A only

    B. B only

    C. Both A and B

    D. Neither A nor B

35. Which of the following steps is a technician LEAST LIKELY to perform when pressing the wrist pin into the piston and connecting rod?

    A. Align the bores in the piston and connecting rod.

    B. Heat the small end of the rod.

    C. Make sure the position marks on the piston and connecting rod are oriented properly.

    D. Heat the wrist pin.

36. Technician A says if the free length of the valve spring is within specifications, it does not need to be tension tested. Technician B says a valve spring that is not square may cause uneven valve seat wear. Who is correct?

    A. A only
    B. B only
    C. Both A and B
    D. Neither A nor B

37. Technician A says the exhaust lobes on a flat tappet lifter camshaft may wear more severely than the intake lobes. Technician B says this is due to increased pressure during valve opening. Who is correct?

    A. A only
    B. B only
    C. Both A and B
    D. Neither A nor B

38. Technician A says the crankshaft keyway should be inspected for chipping and wear. Technician B says a crankshaft that is cracked may be welded and reused. Who is correct?

    A. A only
    B. B only
    C. Both A and B
    D. Neither A nor B

39. Technician A says that the lobe face on a flat tappet lifter camshaft is flat. Technician B says a flat tappet lifter will have a concave face. Who is correct?

    A. A only
    B. B only
    C. Both A and B
    D. Neither A nor B

40. Technician A says that oil galley plugs in the block may be pipe thread. Technician B says the oil galley plugs may be cup plugs. Who is correct?

    A. A only
    B. B only
    C. Both A and B
    D. Neither A nor B

41. Technician A says a timing chain can be checked for stretch by using a timing light. Technician B says a stretched chain may cause intermittent poor performance. Who is correct?

    A. A only
    B. B only
    C. Both A and B
    D. Neither A nor B

42. A thermostat in the cooling system helps prevent:

    A. Overheating.

    B. Overcooling.

    C. Heater core leaks.

    D. Hose deterioration.

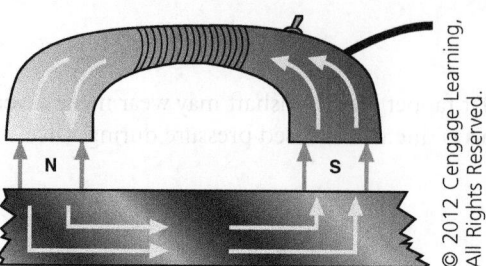

© 2012 Cengage Learning, All Rights Reserved.

43. An electromagnetic-type tester shown in the figure above, along with iron filings, can be used to check for cracks in:

    A. Aluminum intake manifolds.

    B. Pistons.

    C. Aluminum cylinder heads.

    D. Cast iron cylinder heads.

44. The rocker arms on a pushrod engine have a 1.5:1 ratio. This means that:

    A. A cam lift of 0.350 inches will cause the valve to open 0.175 inches.

    B. A cam lift of 0.350 inches will cause the valve to open 0.525 inches.

    C. A cam lift of 0.350 inches will cause the valve to open 0.350 inches.

    D. A cam lift of 0.350 inches will cause the valve to open 0.700 inches.

45. Accessory drive belts are being inspected. Technician A says the automatic tensioner should not vibrate when the engine is running without the A/C engaged. Technician B says all pulleys driven by a belt must be lined up. Who is correct?

    A. A only

    B. B only

    C. Both A and B

    D. Neither A nor B

46. The cylinders in a block have a pronounced ridge at the top. Technician A says this could be caused by excess fuel in the combustion chamber. Technician B says this could be caused by ring gap that was too tight. Who is correct?

    A. A only

    B. B only

    C. Both A and B

    D. Neither A nor B

47. Technician A says supercharger lubrication is done by the engine lubrication system. Technician B says supercharger oil is separate and should be changed according to manufacturer's specifications. Who is correct?

    A. A only
    B. B only
    C. Both A and B
    D. Neither A nor B

48. Technician A says in-block camshaft bores do not warp so they will not have to be checked. Technician B says all in-block camshaft bearings are usually the same size. Who is correct?

    A. A only
    B. B only
    C. Both A and B
    D. Neither A nor B

49. Technician A says the oil pump may be driven by the camshaft through the distributor. Technician B says the oil pump may be driven by the crankshaft. Who is correct?

    A. A only
    B. B only
    C. Both A and B
    D. Neither A nor B

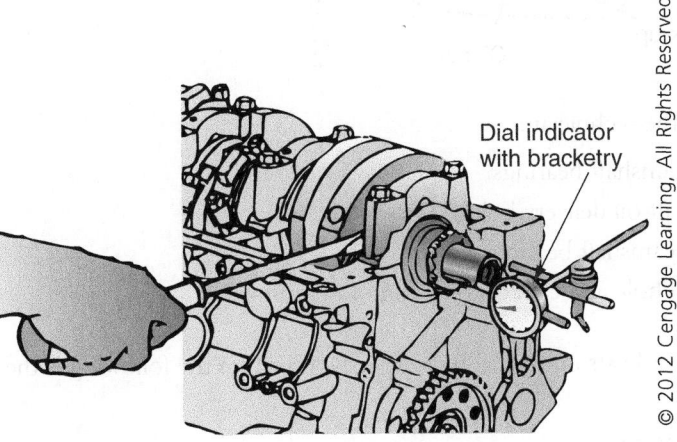

Dial indicator with bracketry

© 2012 Cengage Learning, All Rights Reserved.

50. In the figure above, the technician is:

    A. Measuring main bearing clearance.
    B. Measuring rod bearing clearance.
    C. Checking for crankshaft warp.
    D. Checking thrust bearing clearance.

PREPARATION EXAM 5

1. The customer complains of a loud noise coming from the engine compartment. Which of the following would be the LEAST LIKELY to cause the noise?

   A. Cracked exhaust manifold

   B. A vacuum leak

   C. A cracked flexplate

   D. Carbon buildup on the pistons

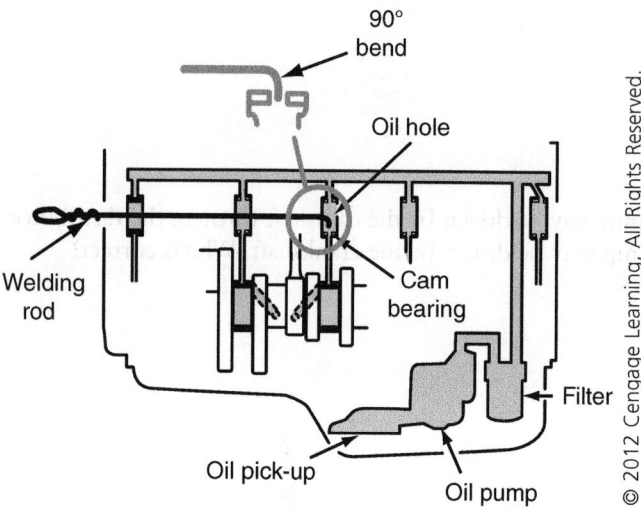

2. The figure above shows the technician:

   A. Installing in-block camshaft bearings.

   B. Cleaning the camshaft oil delivery holes.

   C. Removing in-block camshaft bearings.

   D. Checking journal oil hole and cam bearing oil hole alignment.

3. A high-mileage vehicle overheats only at highway speeds. Which of the following is the most likely to be the cause?

   A. A stuck closed thermostat

   B. A defective radiator cap vacuum valve

   C. An inoperative cooling fan

   D. A radiator whose core is clogged with deposits

4. A battery rated at 600 cold cranking amps (CCA) is load tested at one-half of its rated CCA for 15 seconds. The results show 10.1 volts. This indicates the battery:

   A. Is bad and should be replaced.

   B. Needs recharging.

   C. Is good.

   D. Should be retested for 30 seconds at load.

5. Technician A says a full floating piston pin will ride in a steel bushing in the small end of the connecting rod. Technician B says a bronze bushing is used. Who is correct?

    A.   A only
    B.   B only
    C.   Both A and B
    D.   Neither A nor B

6. Technician A says that o-ring type valve stem seals are installed after the spring and retainer are installed. Technician B says that positive lock valve stem seals ride on the valve stem. Who is correct?

    A.   A only
    B.   B only
    C.   Both A and B
    D.   Neither A nor B

7. A vehicle is brought in with excessive oil loss. Technician A says the loss may be through the rear main seal. Technician B says bad valve stem seals can cause excessive oil consumption. Who is correct?

    A.   A only
    B.   B only
    C.   Both A and B
    D.   Neither A nor B

8. Technician A says that piston ring gap that is too wide could cause ring binding and breakage. Technician B says that piston ring gap that is too tight would cause combustion gas blowby. Who is correct?

    A.   A only
    B.   B only
    C.   Both A and B
    D.   Neither A nor B

9. A customer brings his car to you complaining of a loss of oil pressure after extended driving. Which of the following is the most likely cause?

    A.   Worn main bearings
    B.   Weak oil pump
    C.   Defective oil sending unit
    D.   Trash in the oil pan stopping-up the oil pickup screen

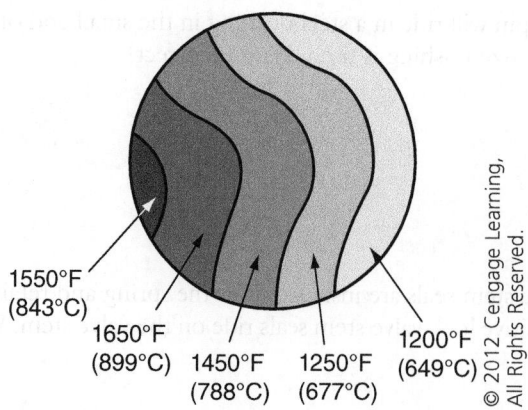

1550°F
(843°C)

1650°F
(899°C)   1450°F   1250°F   1200°F
          (788°C)   (677°C)   (649°C)

© 2012 Cengage Learning, All Rights Reserved.

10. Technician A says the figure above shows the valve head temperatures for a properly seated valve. Technician B says the figure shows the valve head temperatures for a valve that is not properly contacting the seat. Who is correct?

    A. A only
    B. B only
    C. Both A and B
    D. Neither A nor B

11. An engine cranks over very slowly and may not start. Which is the LEAST LIKELY cause?

    A. Excessive electrical resistance at the starter
    B. Weak battery
    C. Broken timing belt
    D. Worn ignition switch contacts

12. Technician A says that all front-wheel drive engines can be removed through the top of the engine compartment. Technician B says that some front-wheel drive vehicle engines must be removed from underneath the vehicle. Who is correct?

    A. A only
    B. B only
    C. Both A and B
    D. Neither A nor B

13. A customer's vehicle is leaking oil on his driveway from the rear of the engine. Technician A says the leak may be coming from the rear main seal. Technician B says if a visual inspection does not pinpoint the leak, florescent dye should be put in the oil. Who is correct?

    A. A only
    B. B only
    C. Both A and B
    D. Neither A nor B

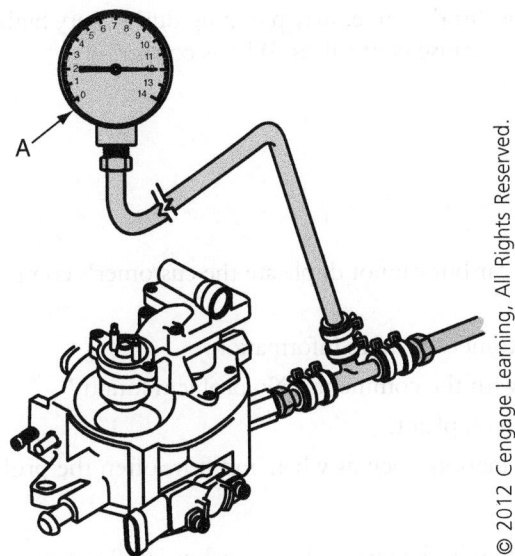

© 2012 Cengage Learning, All Rights Reserved.

14. In the figure above, the technician is checking:

    A. Oil pressure.

    B. Engine vacuum.

    C. Fuel pressure.

    D. The PCV system.

15. During a vacuum check of a poorly running engine, it is found that vacuum holds steady between 8 and 14 in. Hg. Technician A says this may be caused by weak valve springs. Technician B says ignition timing that is off could cause this problem. Who is correct?

    A. A only

    B. B only

    C. Both A and B

    D. Neither A nor B

16. While checking coolant level in the radiator, red oily specks are noted floating on the coolant. Technician A says this is rust and the system should be flushed. Technician B says this is probably transmission fluid and the cooler built into the radiator is leaking. Who is correct?

    A. A only

    B. B only

    C. Both A and B

    D. Neither A nor B

17. The technician has done a compression test on an engine and found cylinders #1 and #3 to have low compression. What would be the next test he should do in his diagnosis?

    A. Power balance test

    B. Cylinder leakage test

    C. Wet compression test

    D. Dynamic compression test

18. Technician A says a hydraulic lifter, in a pushrod engine, may pump up during very high RPM operation. Technician B says this may cause bent valves. Who is correct?

     A. A only

     B. B only

     C. Both A and B

     D. Neither A nor B

19. The technician test drives the customer's car but cannot duplicate the customer's complaint. His next step should be:

     A. Ask the service writer to call the customer for more information.

     B. Return the vehicle to the customer with the comment, "No problem found."

     C. Do a repair based on the customer's complaint.

     D. Call the customer to get more information, such as when and how often the problem occurs.

20. Technician A says all piston pins are centered in the piston skirts. Technician B says some piston pins might be offset. Who is correct?

     A. A only

     B. B only

     C. Both A and B

     D. Neither A nor B

21. A cranking compression test on a 4-cylinder engine has been done and the results were:

CYL #1	CYL #2	CYL #3	CYL #4
135 psi	20 psi	20 psi	130 psi

     Technician A says a wet test should be done on cylinders #2 and #3. Technician B says the results indicate a blown head gasket between cylinders #2 and #3. Who is correct?

     A. A only

     B. B only

     C. Both A and B

     D. Neither A nor B

22. A serpentine belt has just been replaced; it squeals in the morning and during acceleration. Technician A says this could be caused by worn pulley grooves. Technician B says a weak tensioner could cause the problem. Who is correct?

     A. A only

     B. B only

     C. Both A and B

     D. Neither A nor B

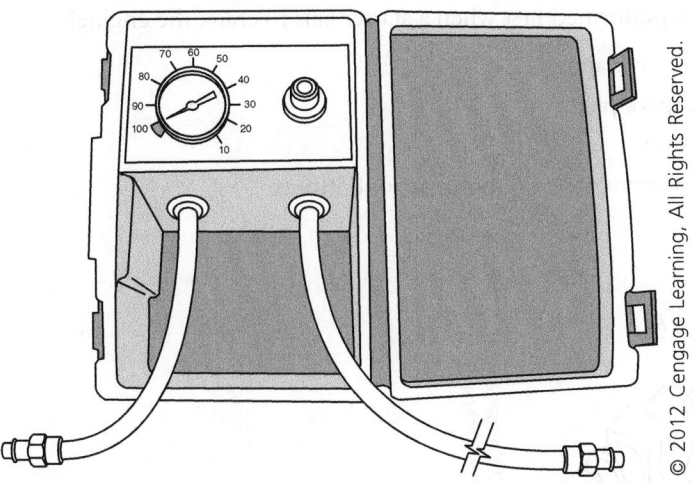

© 2012 Cengage Learning, All Rights Reserved.

23. The gauge set shown in the figure above is used to perform:

    A.   Oil pressure tests.
    B.   Compression tests.
    C.   Cylinder leakage tests.
    D.   Vacuum tests.

24. Technician A says RTV can be used to seal a throttle body injection assembly to the intake manifold. Technician B says RTV is a gasket maker and should not be applied to another gasket. Who is correct?

    A.   A only
    B.   B only
    C.   Both A and B
    D.   Neither A nor B

25. The exhaust on a customer's car is white and has a sweet smell. Technician A says this indicates a rich condition in the cylinder. Technician B says this indicates oil being burned in the cylinder. Who is correct?

    A.   A only
    B.   B only
    C.   Both A and B
    D.   Neither A nor B

26. Technician A says most newly designed engines use low tension piston rings. Technician B says that low tension piston rings decrease friction in the cylinder and help improve fuel economy. Who is correct?

    A.   A only
    B.   B only
    C.   Both A and B
    D.   Neither A nor B

27. Which of these should be performed first when a starter fails to crank the engine?

  A. Measure static battery voltage.

  B. Remove and check spark plugs.

  C. Check for fuel pressure.

  D. Bypass the starter solenoid with a remote starter button.

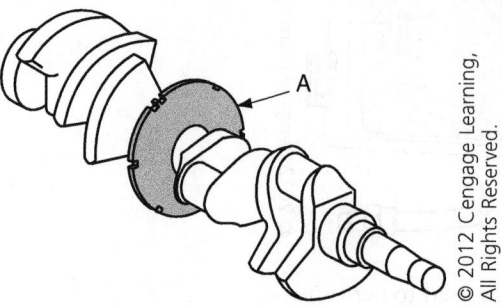

© 2012 Cengage Learning, All Rights Reserved.

28. In the figure above, Technician A says that A is a counter balance weight. Technician B says it is a reluctor ring. Who is correct?

  A. A only

  B. B only

  C. Both A and B

  D. Neither A nor B

29. All of the following are causes of low engine oil pressure EXCEPT:

  A. Worn crankshaft main bearings.

  B. Weak oil pump pressure relief valve spring.

  C. Stuck closed pressure relief valve.

  D. Restricted pushrod oil passages.

30. Technician A says a thermostat rated 180 degrees is fully open at that temperature. Technician B says the thermostat begins to open at 180 degrees. Who is correct?

  A. A only

  B. B only

  C. Both A and B

  D. Neither A nor B

31. Technician A says if a timing belt is removed and is going to be reused, the direction of rotation must be marked. Technician B says a timing belt that has been contaminated with oil can be cleaned and reused. Who is correct?

  A. A only

  B. B only

  C. Both A and B

  D. Neither A nor B

32. Technician A says old coolant is not hazardous and can be disposed of in the shop drain. Technician B says old coolant should be disposed of in an environmentally safe manner. Who is correct?

    A. A only
    B. B only
    C. Both A and B
    D. Neither A nor B

33. When inspected, the crankshaft bearings that exhibit more wear are those furthest from the oil pump; wear is less noticeable as you get closer to the oil pump. What is the most likely reason for the wear?

    A. Engine overloading at low RPM
    B. Over-tightened accessory belts
    C. Dry starts
    D. A warped crankshaft

34. An engine idles rough and the technician notes a high-pitch whistle from the engine compartment. Which of the following would be the most likely cause?

    A. A slipping fan belt
    B. A leaking exhaust pipe
    C. A defective alternator bearing
    D. A vacuum leak on the intake

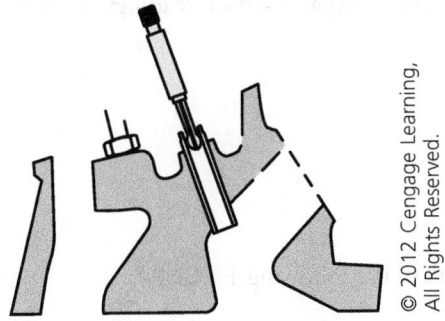

© 2012 Cengage Learning, All Rights Reserved.

35. The figure above shows a technician:

    A. Installing a valve guide.
    B. Measuring stem installed height.
    C. Measuring a valve guide for wear.
    D. Removing a valve guide.

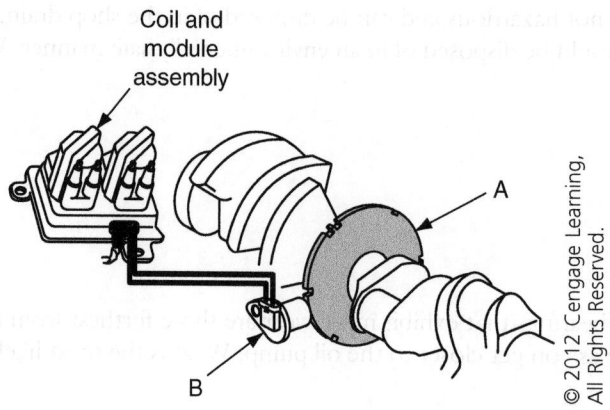

Coil and
module
assembly

A

B

© 2012 Cengage Learning,
All Rights Reserved.

36. In the figure above, Technician A says if the distance between A and B is too wide, it will cause a no-crank condition. Technician B says pickup B is a thermistor. Who is correct?

   A. A only
   B. B only
   C. Both A and B
   D. Neither A nor B

37. Technician A says a cracked exhaust manifold can cause poor fuel economy. Technician B says a cracked exhaust manifold will cause incorrect oxygen sensor readings. Who is correct?

   A. A only
   B. B only
   C. Both A and B
   D. Neither A nor B

38. A defective water pump can be diagnosed by all of the following EXCEPT:

   A. Using a pressure tester.
   B. A coolant leak from the water pump.
   C. A grinding noise from the pump area.
   D. A lower than normal reading on the temperature gauge.

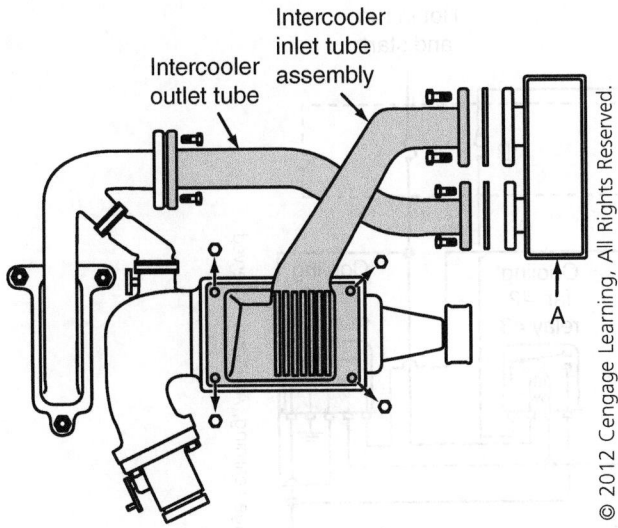

© 2012 Cengage Learning, All Rights Reserved.

39. In the figure above, Technician A says that A is the intercooler. Technician B says a supercharger is driven by exhaust gases. Who is correct?

   A. A only
   B. B only
   C. Both A and B
   D. Neither A nor B

40. All of the following are true about TTY bolts EXCEPT:

   A. Provide a more uniform clamping force.
   B. Usually require special tightening procedures.
   C. Must always be discarded after use.
   D. May be used in connecting rod applications.

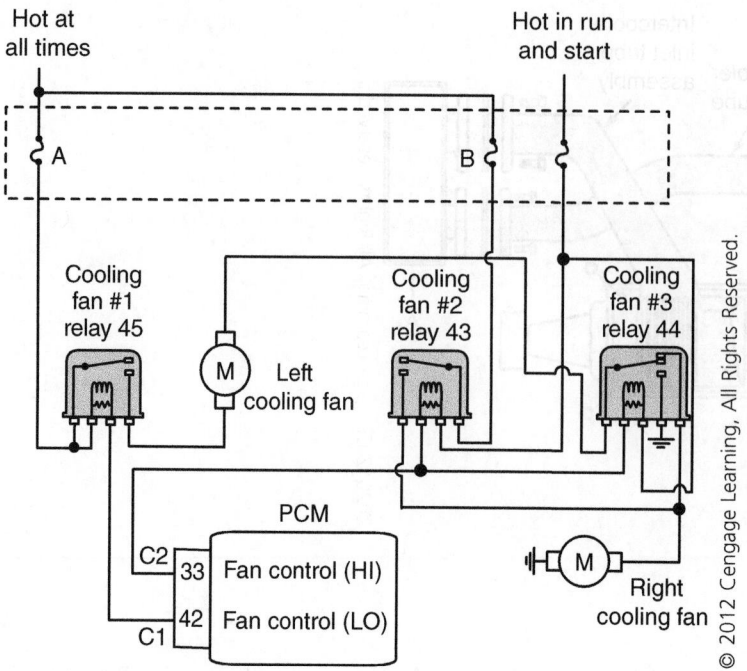

41. In the figure above, Technician A says a blown fuse at point A will keep the left cooling fan from operating. Technician B says a blown fuse at point B will keep the left cooling fan from running in low speed. Who is correct?

    A. A only

    B. B only

    C. Both A and B

    D. Neither A nor B

42. Technician A says that fuel injectors are controlled by the crankshaft position sensor. Technician B says a noid light can be used to check for injector trigger signal. Who is correct?

    A. A only

    B. B only

    C. Both A and B

    D. Neither A nor B

43. Technician A says that valve keepers must be inspected for wear, cracks, and rounded corners. Technician B says that valve keeper lock grooves must also be checked for wear. Who is correct?

    A. A only

    B. B only

    C. Both A and B

    D. Neither A nor B

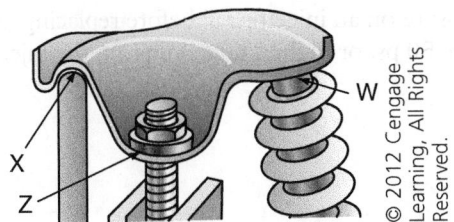

© 2012 Cengage Learning, All Rights Reserved.

44. In the figure above, how is valve lash adjusted?

    A. By adding shims at point W
    B. By adding shims at point X
    C. By rotating nut Z
    D. By replacing the pushrod

45. Technician A says some manufacturers use a sensor to indicate air filter condition. Technician B says air filter condition is displayed on the dash. Who is correct?

    A. A only
    B. B only
    C. Both A and B
    D. Neither A nor B

46. The camshaft bore in an overhead cam engine does not use bearing inserts. Technician A says if the camshaft-to-bore clearance exceeds specifications, bearing inserts could be installed to correct the problem. Technician B says the head must be replaced. Who is correct?

    A. A only
    B. B only
    C. Both A and B
    D. Neither A nor B

47. The LEAST LIKELY cause of camshaft bind would be:

    A. Improperly installed bearings.
    B. Bore misalignment.
    C. Excessive bearing clearance.
    D. Mixed up journal caps.

48. During a power balance test on a port fuel-injected engine, one cylinder is found to have virtually no RPM change. Which of these is the most likely cause?

    A. A faulty crankshaft position sensor
    B. A vacuum leak at the throttle body
    C. A defective plug wire
    D. A faulty camshaft position sensor

49. Technician A says it is important to relieve pressure on an injector rail before replacing an injector. Technician B says fuel pressure may be 50 psi or higher, depending on the injection system. Who is correct?

    A. A only

    B. B only

    C. Both A and B

    D. Neither A nor B

50. After a cold start, the customer notes a loud cracking or popping noise from the engine that quiets as the engine warms up. Technician A says this may be caused by a worn rod bearing. Technician B says this may be normal engine noise caused by lifter bleed down. Who is correct?

    A. A only

    B. B only

    C. Both A and B

    D. Neither A nor B

PREPARATION EXAM 6

1. Technician A says when removing an engine with an automatic transmission, the transmission torque converter is removed with the engine. Technician B says all the accessories must be removed from the engine before removal. Who is correct?

    A. A only

    B. B only

    C. Both A and B

    D. Neither A nor B

2. During teardown, it is noted that several valve stem tips were severely mushroomed. Technician A says the mushroomed tips must be dressed with a file before valve removal. Technician B says this was probably caused by the valve clearance being excessive. Who is correct?

    A. A only

    B. B only

    C. Both A and B

    D. Neither A nor B

3. Technician A says an intercooler may be used to cool the air taken in by the supercharger before it reaches the combustion chamber. Technician B says it is to prevent the supercharger from overheating. Who is correct?

    A. A only

    B. B only

    C. Both A and B

    D. Neither A nor B

4. The technician is investigating an intermittent popping noise from under the hood that occurs when the RPM is raised. Which of the following is the LEAST LIKELY cause for the noise?

   A. Sticky intake valve

   B. Incorrect ignition timing

   C. Broken intake valve spring

   D. Worn rod bearing

5. Technician A says pushrods may be used to deliver oil to the rocker arms. Technician B says a bent pushrod may indicate a valve was struck by a piston. Who is correct?

   A. A only

   B. B only

   C. Both A and B

   D. Neither A nor B

6. The customer questions the large amount of blue smoke coming out of his tailpipe continuously. Technician A says the problem may be worn main bearings. Technician B says this may be caused by excess fuel in the combustion chamber. Who is correct?

   A. A only

   B. B only

   C. Both A and B

   D. Neither A nor B

7. The water pump is being replaced on a rear-wheel drive vehicle with a longitudinally placed engine. What is the LEAST LIKELY part that will have to be removed?

   A. The cooling fan

   B. The fan belt

   C. The radiator

   D. The fan shroud

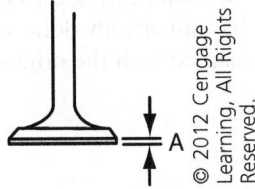

© 2012 Cengage Learning, All Rights Reserved.

8. Technician A says that A in the figure above indicates the valve margin. Technician B says typical valve margin should measure no less than 0.031″. Who is correct?

   A. A only

   B. B only

   C. Both A and B

   D. Neither A nor B

9. A poor performing engine has a cranking compression test done. The test fails with all cylinders showing low compression. The technician did a wet compression test and all cylinders increased substantially. Technician A says the wet compression test results indicate the head gasket is not sealing. Technician B says the wet compression test results indicate cylinder and ring wear. Who is correct?

    A. A only

    B. B only

    C. Both A and B

    D. Neither A nor B

10. An oil pan is being reinstalled on the engine. Technician A says that the oil pan can be sealed using RTV. Technician B says that a new gasket may be installed. Who is correct?

    A. A only

    B. B only

    C. Both A and B

    D. Neither A nor B

11. Technician A says that all coolants are the same if they are ethylene glycol-based. Technician B says different coolant types can be mixed with no problems. Who is correct?

    A. A only

    B. B only

    C. Both A and B

    D. Neither A nor B

12. Which of the following tools would be used to measure valve-to-guide clearance?

    A. A micrometer and feeler gauges

    B. A dial caliper and feeler gauges

    C. A feeler gauge and machinist's rule

    D. A small hole gauge and a micrometer

13. Technician A says before disassembly of the engine block begins, all main bearing caps must be marked for their location and orientation, if the manufacturer has not already done so. Technician B says cracked design connecting rod caps should be marked with the original orientation and cylinder. Who is correct?

    A. A only

    B. B only

    C. Both A and B

    D. Neither A nor B

14. The vehicle has an intermittent no-crank condition; the starter relay has been replaced. Which of the following is the LEAST LIKELY cause?

    A. Loose connection at the battery

    B. Worn neutral safety switch

    C. Open circuit in the battery positive wire on the starter

    D. Worn ignition switch

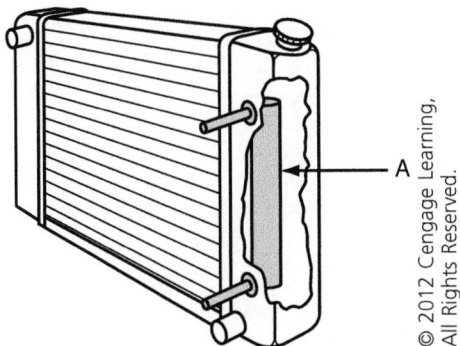

© 2012 Cengage Learning,
All Rights Reserved.

15. In the figure above, Technician A says A is an auxiliary heater. Technician B says A is an automatic transmission fluid cooler. Who is correct?

    A. A only
    B. B only
    C. Both A and B
    D. Neither A nor B

16. Which of the following is the LEAST LIKELY location for an engine oil leak?

    A. Oil pan gasket
    B. Valve cover gasket
    C. Oil pressure sending unit
    D. Upper intake manifold gasket

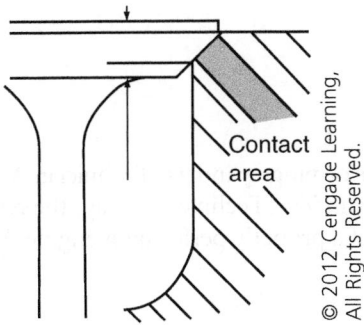

© 2012 Cengage Learning,
All Rights Reserved.

17. In the figure above, the seat-to-valve contact area is too high on the valve face and the seat is too wide. Technician A says to narrow the seat with a 45 degree angle. Technician B says to top cut the seat with a 30 degree angle. Who is correct?

    A. A only
    B. B only
    C. Both A and B
    D. Neither A nor B

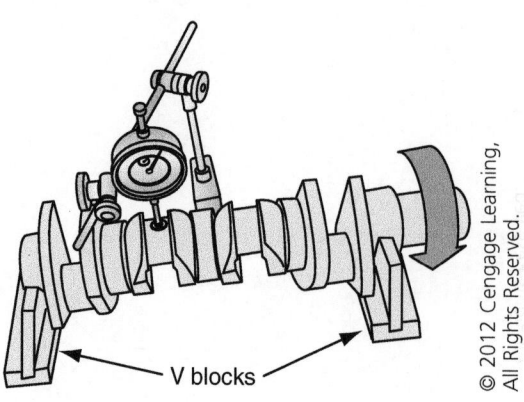
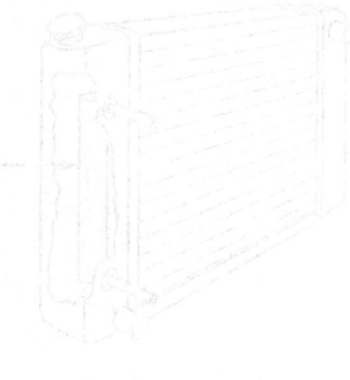

V blocks

© 2012 Cengage Learning, All Rights Reserved.

18. The figure above shows a technician:

    A. Checking journal diameter.
    B. Polishing the journal.
    C. Checking a crankshaft for runout.
    D. Demagnetizing the crankshaft.

19. Technician A says a V-belt that is too tight can wear out water pump bearings. Technician B says that a V-belt that is too tight can cause the #1 crank journal bearing to wear out on the upper half. Who is correct?

    A. A only
    B. B only
    C. Both A and B
    D. Neither A nor B

20. A power balance test is being done to pinpoint low performing cylinders. Technician A says there will be a drop in RPM on the poor performing cylinders. Technician B says the results should be within $+/- 10$ percent across all cylinders of a properly performing engine. Who is correct?

    A. A only
    B. B only
    C. Both A and B
    D. Neither A nor B

21. Technician A says all piston pins are press fit into the connecting rod. Technician B says some piston pins are full floating. Who is correct?

    A. A only
    B. B only
    C. Both A and B
    D. Neither A nor B

22. A cylinder leakage test has been performed and one cylinder failed with a leakage of over 80 percent. Technician A says the leakage could be into the cooling system. Technician B says testing for combustion gases in the radiator would confirm the diagnosis. Who is correct?

    A. A only
    B. B only
    C. Both A and B
    D. Neither A nor B

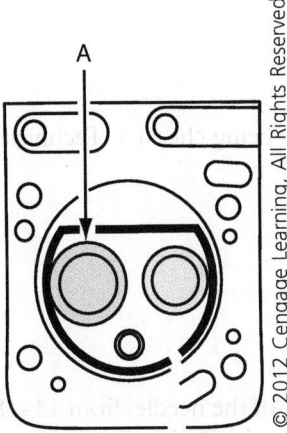

© 2012 Cengage Learning, All Rights Reserved.

23. Technician A says that A in the figure above is an exhaust valve. Technician B says the exhaust valve is always smaller than the intake valve. Who is correct?

    A. A only
    B. B only
    C. Both A and B
    D. Neither A nor B

24. Technician A says that most cylinder wear occurs in the center of piston ring travel. Technician B says that most cylinder wear occurs at the bottom of piston ring travel. Who is correct?

    A. A only
    B. B only
    C. Both A and B
    D. Neither A nor B

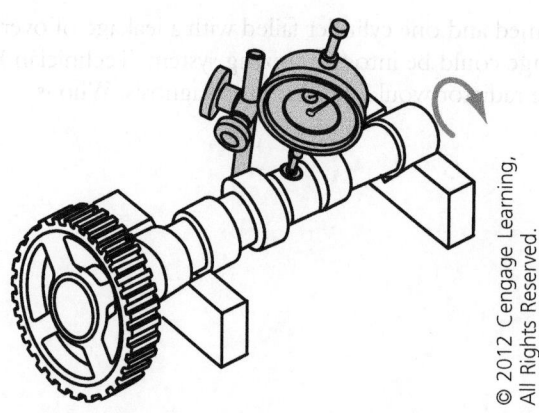

© 2012 Cengage Learning,
All Rights Reserved.

25. In the figure above, Technician A says the camshaft lobe lift is being checked. Technician B says the camshaft is being checked for runout. Who is correct?

    A. A only

    B. B only

    C. Both A and B

    D. Neither A nor B

26. With a vacuum gauge hooked to the engine, rapid fluctuation of the needle, from 14–18 in. Hg, is observed; this increases with RPM. Technician A says the intake manifold is restricted. Technician B says the valve springs may be weak or broken. Who is correct?

    A. A only

    B. B only

    C. Both A and B

    D. Neither A nor B

27. An engine overheats in stop-and-go traffic but does not overheat on the highway. Technician A says a defective radiator cap may be the cause. Technician B says an inoperative electric cooling fan may be the cause. Who is correct?

    A. A only

    B. B only

    C. Both A and B

    D. Neither A nor B

28. The LEAST LIKELY cause of excessive blue smoke from the exhaust of a turbocharged engine is:

    A. Worn piston rings.

    B. Bad valve stem seals.

    C. A PCV valve stuck in the open position.

    D. Worn turbocharger seals.

29. Technician A says that a worn valve guide should be reconditioned or replaced before the valve seats are reconditioned. Technician B says the valve seats should be replaced before repairing the valve guide. Who is correct?

    A. A only
    B. B only
    C. Both A and B
    D. Neither A nor B

30. All of the following are symptoms of a stuck open PCV valve EXCEPT:

    A. Rough engine idle.
    B. A lean air/fuel ratio.
    C. Blowby gases in the air filter.
    D. The engine stalling.

31. An engine miss is being diagnosed using a cylinder leakage test. Technician A says that a 20 percent leakage is acceptable. Technician B says air coming out the intake manifold indicates a cracked cylinder head. Who is correct?

    A. A only
    B. B only
    C. Both A and B
    D. Neither A nor B

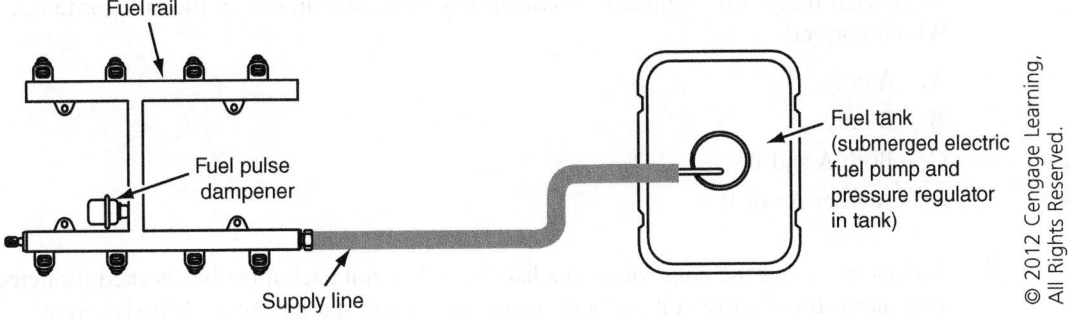

32. Technician A says the figure above shows a returnless fuel system. Technician B says the fuel pump and pressure regulator are in the gas tank. Who is correct?

    A. A only
    B. B only
    C. Both A and B
    D. Neither A nor B

33. Technician A says typical piston-to-bore clearance is 0.020 to 0.030 inch. Technician B says typical piston-to-bore clearance is 0.001 to 0.002 inch. Who is correct?

    A. A only

    B. B only

    C. Both A and B

    D. Neither A nor B

34. Technician A says the air intake duct inlet is in the engine compartment. Technician B says to check for contaminates in the inlet when checking or replacing the air filter. Who is correct?

    A. A only

    B. B only

    C. Both A and B

    D. Neither A nor B

35. Technician A says stuck or sticky valves may cause bent valves. Technician B says a timing belt off by three teeth may cause bent valves. Who is correct?

    A. A only

    B. B only

    C. Both A and B

    D. Neither A nor B

36. Technician A says an engine oil cooler may be located in one of the radiator tanks. Technician B says a transmission oil cooler may be located in one of the radiator tanks. Who is correct?

    A. A only

    B. B only

    C. Both A and B

    D. Neither A nor B

37. Technician A says the plug wires on a failed ignition coil pack must be checked for defects. Technician B says most coil packs are considered waste spark systems. Who is correct?

    A. A only

    B. B only

    C. Both A and B

    D. Neither A nor B

38. Technician A says that when adjusting valves on an engine with mechanical adjustment, the piston must be at BDC on the exhaust stroke. Technician B says some mechanical valve adjustment is done by changing removable lash pads on a bucket. Who is correct?

    A. A only

    B. B only

    C. Both A and B

    D. Neither A nor B

39. The engine cranks but will not start. Which of the following is not a requirement for the engine to start and run?

    A. Compression
    B. Oil pressure
    C. Fuel pressure
    D. Spark

40. A technician begins his diagnostic procedure to repair a vehicle. Which of the following would be the LEAST LIKELY steps he would take?

    A. Road test the vehicle.
    B. Question the customer for more information regarding the problem.
    C. Make sure the original complaint is fixed.
    D. Start with the most difficult tests.

41. Technician A says the main bearing bores are checked for alignment with a straightedge and a feeler gauge. Technician B says main bearing bores that are out of alignment can be corrected by line boring. Who is correct?

    A. A only
    B. B only
    C. Both A and B
    D. Neither A nor B

42. Technician A says the upper oil galleys in the block may be sealed with pipe plugs. Technician B says the oil galley plugs must be removed to properly clean the galleys. Who is correct?

    A. A only
    B. B only
    C. Both A and B
    D. Neither A nor B

43. Technician A says that valve train related noise could be caused by a low oil level. Technician B says that valve train related noise could be caused by valves that need to be adjusted. Who is correct?

    A. A only
    B. B only
    C. Both A and B
    D. Neither A nor B

44. A technician is changing oil in a customer's vehicle and needs to know what weight oil to put in. Technician A says the weight of oil does not matter as long as it is multigrade oil. Technician B says the technician should look on the oil filler cap or in the owner's manual. Who is correct?

    A. A only
    B. B only
    C. Both A and B
    D. Neither A nor B

45. Technician A says that when removing a cylinder head from an OHC engine, the camshaft may have to be removed first. Technician B says the cylinder head should be cold before removal. Who is correct?

    A. A only
    B. B only
    C. Both A and B
    D. Neither A nor B

46. The customer is concerned about his oil pressure gauge dropping after the vehicle is driven for about 20 miles, although when he starts up, it is normal. When the dipstick is checked after shutdown, it shows below the add mark. When the dipstick is checked after sitting, the oil level shows full. Which of the following is the most likely cause?

    A. Worn camshaft bearings
    B. Leaking rear main seal
    C. Blocked oil return holes in the head
    D. Weak oil pump

47. Technician A says that during engine reassembly, all rod bearing clearances should be checked. Technician B says the rod bearing inserts should extend slightly above the rod and cap; this is called bearing crush. Who is correct?

    A. A only
    B. B only
    C. Both A and B
    D. Neither A nor B

48. While performing a cranking compression test on a 4-cylinder engine, the technician notes that one cylinder has a pressure reading of 60 psi, 44 kPa, while the others have a reading of 135 psi, 931 kPa. Technician A says performing a cylinder leakage test will indicate where the pressure is leaking. Technician B says the vehicle has worn valve guides. Who is correct?

    A. A only
    B. B only
    C. Both A and B
    D. Neither A nor B

49. Which of the following could cause symptoms of overheating even though the engine temperature is normal?

    A. A stuck closed thermostat
    B. A defective radiator cap
    C. A defective temperature sending unit
    D. A missing thermostat

50. The customer complains of a loud thump when he accelerates from a stop. Technician A says it could be caused by worn crankshaft thrust bearings. Technician B says it could be caused by a cracked flexplate. Who is correct?

    A. A only
    B. B only
    C. Both A and B
    D. Neither A nor B

Answer Keys and Explanations

INTRODUCTION

Included in this section are the answer keys for each preparation exam, followed by individual, detailed answer explanations and a reference identifying the designated task area being assessed by each specific question. This additional reference information may prove useful if you need to refer back to the task list located in section 4 of this book for additional support.

PREPARATION EXAM 1—ANSWER KEY

1.	C	21.	B	41.	D
2.	A	22.	A	42.	C
3.	C	23.	A	43.	D
4.	B	24.	B	44.	D
5.	B	25.	D	45.	C
6.	C	26.	D	46.	C
7.	D	27.	C	47.	B
8.	A	28.	D	48.	D
9.	C	29.	A	49.	B
10.	A	30.	C	50.	A
11.	C	31.	C		
12.	D	32.	C		
13.	A	33.	B		
14.	B	34.	D		
15.	A	35.	B		
16.	B	36.	A		
17.	A	37.	B		
18.	C	38.	C		
19.	A	39.	A		
20.	D	40.	D		

PREPARATION EXAM 1—EXPLANATIONS

TASK A.1

1. Technician A says that it is important to test drive the customer's vehicle to verify the concern as a part of the diagnosis procedure. Technician B says it is always a good idea to check for associated technical service bulletins. Who is correct?

    A. A only
    B. B only
    C. Both A and B
    D. Neither A nor B

    Answer A is incorrect. Technician B is also correct.

    Answer B is incorrect. Technician A is also correct.

    Answer C is correct. Both Technicians are correct. A test drive lets you confirm the customer complaint, and checking for technical service bulletins may save diagnostic time.

    Answer D is incorrect. Both Technicians are correct.

TASK C.7

2. Technician A says main bearing oil clearance can be checked with Plastigauge®. Technician B says main bearing oil clearance can be checked with a feeler gauge. Who is correct?

    A. A only
    B. B only
    C. Both A and B
    D. Neither A nor B

    Answer A is correct. Only Technician A is correct. Main bearing oil clearance can be checked with Plastigauge® by placing the Plastigauge® across the journal with the bearings installed and torqueing the cap. Remove the cap and use the Plastigauge® scale; record the clearance.

    Answer B is incorrect. A feeler gauge is not a tool to check main bearing oil clearance. It will not conform to the journal to give an accurate reading.

    Answer C is incorrect. Only Technician A is correct.

    Answer D is incorrect. Technician A is correct.

TASK E.1

3. Technician A says that when the fuel lines are reconnected on a replaced engine, the o-ring seals on the fuel lines must be replaced. Technician B says the injector o-rings should be replaced when they are reinstalled. Who is correct?

    A. A only
    B. B only
    C. Both A and B
    D. Neither A nor B

    Answer A is incorrect. Technician B is also correct.

    Answer B is incorrect. Technician A is also correct.

    Answer C is correct. Both Technicians are correct. Both the fuel line seals and injector seals must be replaced to ensure there is no fuel leakage.

    Answer D is incorrect. Both Technicians are correct.

4. The oil filter on a customer's car is bulged out. Technician A says this could be caused by improper filter installation. Technician B says this could be caused by a stuck closed pressure regulator. Who is correct?

TASK D.2

   A. A only

   B. B only

   C. Both A and B

   D. Neither A nor B

   Answer A is incorrect. Improper installation would cause a severe oil leak.

   Answer B is correct. Only Technician B is correct. If the pressure relief valve in the oil pump sticks, oil pressure will go above allowable specifications and bulge or blow off the oil filter.

   Answer C is incorrect. Only Technician B is correct.

   Answer D is incorrect. Technician B is correct.

5. A cylinder head for an OHC engine is being inspected. The deck warpage is greater than the maximum allowed. Technician A says the head must be replaced. Technician B says the head might be able to be straightened, trued, and reused. Who is correct?

TASK B.2

   A. A only

   B. B only

   C. Both A and B

   D. Neither A nor B

   Answer A is incorrect. The head can be straightened and resurfaced.

   Answer B is correct. Only Technician B is correct. A warped head can be repaired by straightening and resurfacing.

   Answer C is incorrect. Only Technician B is correct.

   Answer D is incorrect. Technician B is correct.

6. Technician A says that a turbocharger's boost is controlled by a wastegate. Technician B says the wastegate is controlled by engine vacuum. Who is correct?

TASK E.3

   A. A only

   B. B only

   C. Both A and B

   D. Neither A nor B

   Answer A is incorrect. Technician B is also correct.

   Answer B is incorrect. Technician A is also correct.

   Answer C is correct. Both Technicians are correct. The amount of turbo boost is controlled by a wastegate that is opened or closed based on engine vacuum.

   Answer D is incorrect. Both Technicians are correct.

TASK D.1

7. All of the following are causes of low engine oil pressure EXCEPT:

    A.   Worn camshaft bearings.

    B.   Partially plugged oil pickup screen.

    C.   Worn crankshaft bearings.

    D.   Restricted pushrod passages.

Answer A is incorrect. Worn camshaft bearings will cause low engine oil pressure as there will be more oil runoff.

Answer B is incorrect. A partially plugged pickup screen will restrict oil flow through the pump resulting in low pressure.

Answer C is incorrect. Worn crankshaft bearings will allow oil runoff resulting in low oil pressure.

Answer D is correct. Restricted pushrod passages will only prevent oil from reaching the rocker arms and will not result in low oil pressure.

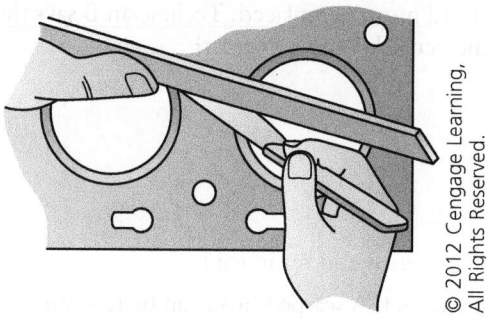

© 2012 Cengage Learning, All Rights Reserved.

TASK B.2

8. In the figure above, what is the technician checking?

    A.   Head warpage

    B.   Piston protrusion

    C.   Surface finish

    D.   Head bolt hole alignment

Answer A is correct. Using a straightedge and feeler gauge, the technician is checking for warpage.

Answer B is incorrect. Piston protrusion is measured with a depth gauge.

Answer C is incorrect. Surface finish is checked with a profilometer.

Answer D is incorrect. Head bolt alignment would require the head to be on the block.

9. The customer is concerned about poor performance at highway speeds. The technician talks to the owner about the car's repair history and is told that the timing belt, water pump, and crank seal have recently been replaced, and the problem started afterward. What is the most likely cause for poor performance?

TASK B.13

    A. The serpentine belt is loose.

    B. The sparkplug wires were misrouted.

    C. The camshaft and crankshaft are out of sync.

    D. The catalytic converter is restricted.

Answer A is incorrect. A loose serpentine belt will cause a squeal but not affect performance.

Answer B is incorrect. Misrouted sparkplug wires would cause a misfire and poor performance all the time.

Answer C is correct. If the camshaft and crankshaft were not positioned properly during timing, belt replacement performance would suffer at higher engine RPM.

Answer D is incorrect. A restricted convertor would cause performance problems at any speed.

10. Technician A says the air intake hose assembly may cause poor performance if it is not properly attached. Technician B says all vehicles use a map sensor to tell the electronic control module (ECM) the amount of ingested air. Who is correct?

TASK E.2

    A. A only

    B. B only

    C. Both A and B

    D. Neither A nor B

Answer A is correct. Only Technician A is correct. If the air intake hose assembly is poorly attached, it will allow unmetered air into the engine, which may cause stumbling and poor performance.

Answer B is incorrect. Map sensors are only one of the ways to tell the ECM the volume of air flow into the intake. Many manufacturers use an air flow sensor in the air intake ductwork.

Answer C is incorrect. Only Technician A is correct.

Answer D is incorrect. Technician A is correct.

11. The customer's complaint is that there is blue smoke emitting from the tailpipe on initial startup in the morning, which seems to dissipate quickly. Technician A says that the blue smoke is caused by oil burning in the combustion chamber. Technician B says the oil probably is entering the combustion chamber through deteriorated valve stem seals. Who is correct?

TASK A.5

    A. A only

    B. B only

    C. Both A and B

    D. Neither A nor B

Answer A is incorrect. Technician B is also correct.

Answer B is incorrect. Technician A is also correct.

Answer C is correct. Both Technicians are correct. Blue smoke that quickly dissipates on initial cold start is a sign that the valve stem seals are allowing oil to run down the valve stem into the combustion chamber after warm shutdown.

Answer D is incorrect. Both technicians are correct.

TASK B.11

12. All lifters in an overhead valve engine are cupped (concave). Technician A says the lifters can be replaced with no further parts replacement. Technician B says that they can be replaced with roller lifters for longer life. Who is correct?

    A. A only

    B. B only

    C. Both A and B

    D. Neither A nor B

Answer A is incorrect. A worn lifter indicates a worn camshaft lobe. If the lifter is replaced without replacing the camshaft, it will fail in a short time.

Answer B is incorrect. A roller lifter cannot be used with a flat tappet camshaft. The lobe on a flat tappet cam is slightly tapered to cause the lifter to spin; a roller camshaft is flat across the lobe.

Answer C is incorrect. Neither Technician is correct.

Answer D is correct. Neither Technician is correct. A camshaft and flat tappet lifters wear to mate during break-in and normal use. If one is replaced, it is recommended to replace both components. Roller lifters would only contact the flat tappet cam in one small area. They are not made for use with the profile of a flat tappet camshaft.

TASK E.5

13. Technician A says a PCV valve hose that is restricted could cause oil accumulation in the air cleaner. Technician B says a PCV valve hose that is restricted will cause a stumble on acceleration. Who is correct?

    A. A only

    B. B only

    C. Both A and B

    D. Neither A nor B

Answer A is correct. Only Technician A is correct. A restricted PCV valve will cause increased crankcase pressure and may cause oil accumulation in the air cleaner.

Answer B is incorrect. A restricted PCV hose will not cause a stumble. A broken hose could cause a stumble, since unmetered air is being introduced into the intake manifold.

Answer C is incorrect. Only Technician A is correct.

Answer D is incorrect. Technician A is correct.

TASK C.1

14. Technician A says that during engine removal, it is okay to open the hoses on the A/C compressor to relieve pressure so you can remove the compressor. Technician B says that the refrigerant in an A/C system must be recovered into an A/C recovery machine. Who is correct?

    A. A only

    B. B only

    C. Both A and B

    D. Neither A nor B

Answer A is incorrect. It is never permissible to vent refrigerant into the atmosphere.

Answer B is correct. Only Technician B is correct. All refrigerant must be recovered into a refrigerant recovery machine.

Answer C is incorrect. Only Technician B is correct.

Answer D is incorrect. Technician B is correct.

15. A cylinder head has been removed and the technician is inspecting the head gasket. He finds the fire ring on cylinder #2 has broken. Technician A says this could be caused by detonation in the cylinder. Technician B says this could be caused by the wrong oil being used. Who is correct?

TASK B.2

    A.   A only

    B.   B only

    C.   Both A and B

    D.   Neither A nor B

Answer A is correct. Only Technician A is correct. Detonation in the cylinder can cause the fire ring to flex excessively and break.

Answer B is incorrect. The wrong oil will not affect the fire ring. However, it may affect bearing wear.

Answer C is incorrect. Only Technician A is correct.

Answer D is incorrect. Technician A is correct.

16. The customer's complaint is that the engine cranks over quickly but does not start. Technician A says it could be caused by the battery. Technician B says it could be caused by a broken timing belt. Who is correct?

TASK A.2

    A.   A only

    B.   B only

    C.   Both A and B

    D.   Neither A nor B

Answer A is incorrect. A weak battery would cause slow cranking.

Answer B is correct. Only Technician B is correct. A broken timing belt will cause the engine to spin over rapidly.

Answer C is incorrect. Only Technician B is correct.

Answer D is incorrect. Technician B is correct.

17. Technician A says corrosion on the battery terminals may cause slow cranking speed. Technician B says high resistance in the starter circuit may result in low cranking speed and high current draw. Who is correct?

TASK E.4

    A.   A only

    B.   B only

    C.   Both A and B

    D.   Neither A nor B

Answer A is correct. Only Technician A is correct. Corrosion on the battery terminals will cause high resistance and a slow starter cranking speed.

Answer B is incorrect. High resistance in the starter circuit may result in low cranking speeds but will result in low current draw.

Answer C is incorrect. Only Technician A is correct.

Answer D is incorrect. Technician A is correct.

TASK B.17

18. Technician A says that when head gaskets are installed, it is important to read the orientation instructions printed on the head gasket. Technician B says an improperly oriented head gasket may restrict coolant flow to the heads. Who is correct?

   A.   A only

   B.   B only

   C.   Both A and B

   D.   Neither A nor B

   Answer A is incorrect. Technician B is also correct.

   Answer B is incorrect. Technician A is also correct.

   Answer C is correct. Both Technicians are correct. A head gasket that has a specified orientation will have instructions printed on the gasket. A head gasket that is reversed, or put on the wrong side of a V-engine block, may cover up coolant or oil passages.

   Answer D is incorrect. Both Technicians are correct.

TASK C.16

19. Technician A says RTV can be used to seal the oil pan in place of a gasket. Technician B says RTV can be used on a head gasket. Who is correct?

   A.   A only

   B.   B only

   C.   Both A and B

   D.   Neither A nor B

   Answer A is correct. Only Technician A is correct. RTV is a gasket maker and is suitable for sealing an oil pan in place of a gasket.

   Answer B is incorrect. A head gasket should have no sealer applied. RTV could cause the head gasket to fail.

   Answer C is incorrect. Only Technician A is correct.

   Answer D is incorrect. Technician A is correct.

TASK E.7

20. A properly working catalytic converter converts HC, CO, and NOx into:

   A. O_3, H_2O, and NO

   B. H_2O, CO_2, and NO

   C. H_2O, NO, and N_2

   D. H_2O, CO_2, and N_2

   Answer A is incorrect. A converter does not convert HC, CO, and NOx into O_3, H_2O, and NO.

   Answer B is incorrect. A converter does not convert HC, CO, and NOx into H_2O, CO_2, and NO.

   Answer C is incorrect. A converter does not convert HC, CO, and NOx into H_2O, NO, and N_2.

   Answer D is correct. A properly working catalytic converter converts HC into H_2O, CO into CO_2, and NOx into N_2.

21. The customer complains of a thump and vibration when accelerating from a stop. Which of the following is the most likely cause for the condition?

    A.   Out of balance tire

    B.   Broken or weak motor mount

    C.   Spark plug misfire

    D.   A warped brake rotor

Answer A is incorrect. A vibration from an out of balance tire would be felt while the car is moving, not when accelerating from a stop.

Answer B is correct. A broken or weak motor mount will allow the engine to shift under acceleration, causing a thump and a vibration when accelerating from a stop.

Answer C is incorrect. A spark plug misfire can cause a vibration but not a thump on acceleration.

Answer D is incorrect. A warped brake rotor will cause a vibration when stopping, not when starting.

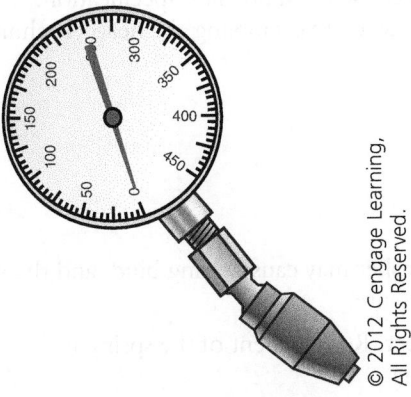

© 2012 Cengage Learning, All Rights Reserved.

22. The gauge in the figure above is used when performing a:

    A.   Compression test.

    B.   Cylinder power balance test.

    C.   Cylinder leakage test.

    D.   Vacuum test.

Answer A is correct. A compression gauge is shown.

Answer B is incorrect. A cylinder power balancing test is usually done with an engine analyzer.

Answer C is incorrect. A cylinder leakage tester is a different device.

Answer D is incorrect. A vacuum test is done with a vacuum gauge.

TASK C.3

23. A head bolt has broken flush with the block deck surface during head removal. Technician A says the remaining bolt can be drilled and a bolt extractor can be used to remove the bolt. Technician B says the broken bolt may be completely drilled out oversize and an oversize head bolt installed. Who is correct?

    A. A only

    B. B only

    C. Both A and B

    D. Neither A nor B

    Answer A is correct. Only Technician A is correct. Drilling the broken bolt with the proper drill size and using a bolt extractor to remove the bolt is the best option.

    Answer B is incorrect. Installing a larger diameter head bolt would require drilling the head hole larger and would create a difference in clamping force in that area of the head.

    Answer C is incorrect. Only Technician A is correct.

    Answer D is incorrect. Technician A is correct.

TASK B.5

24. The valve spring is being checked and fails to meet the open pressure specification. Technician A says to install a valve spring washer when reassembling the head. Technician B says to replace the spring. Who is correct?

    A. A only

    B. B only

    C. Both A and B

    D. Neither A nor B

    Answer A is incorrect. Installing a valve spring washer may cause spring bind, and the washer will increase closed pressure along with open pressure.

    Answer B is correct. Only Technician B is correct. Replacement of the spring is recommended if it fails to meet the specified value.

    Answer C is incorrect. Only Technician B is correct.

    Answer D is incorrect. Technician B is correct.

TASK E.7

25. Technician A says a catalytic converter may be legally removed on a vehicle that is more that 15 years old. Technician B says the catalytic converter significantly restricts exhaust flow. Who is correct?

    A. A only

    B. B only

    C. Both A and B

    D. Neither A nor B

    Answer A is incorrect. A vehicle that came equipped with a catalytic converter may not have the converter removed except to replace it with a new converter under penalty of law.

    Answer B is incorrect. The engine management system is tuned to the exhaust flow on a converter equipped vehicle. The restriction in exhaust flow, although the restriction is minimal, is taken into account during design of the exhaust system.

    Answer C is incorrect. Neither Technician is correct.

    Answer D is correct. Neither Technician is correct. A catalytic converter is part of the emissions system of the vehicle and may not be legally removed without being replaced with a new one at any time during the vehicle's service life. The engine is tuned to perform properly with the minimal restriction of the converter and exhaust modification may cause poor engine performance.

26. Technician A says a serpentine belt that has any cracks in it must be replaced. Technician B says that serpentine belts must be retightened after break-in. Who is correct?

    A. A only

    B. B only

    C. Both A and B

    D. Neither A nor B

TASK D.7

    Answer A is incorrect. Small cracks in the belt are normal unless there are three cracks within 1" of belt length.

    Answer B is incorrect. Serpentine belts do not have a break-in period like a V-belt. No retensioning is necessary.

    Answer C is incorrect. Neither Technician is correct.

    Answer D is correct. Neither Technician is correct. A serpentine belt with three cracks within 1" of belt length must be replaced; small cracks not meeting this guideline do not require belt replacement. A V-belt will stretch slightly during break-in and need retensioning, whereas a serpentine belt will not stretch and will not require retensioning.

27. Florescent dye has been added to the crankcase to help locate an oil leak. The dye will glow when it is exposed to:

    A. A strobe light.

    B. An infrared light.

    C. A blacklight.

    D. A blue light.

TASK A.3

    Answer A is incorrect. A strobe light will not cause the dye to glow.

    Answer B is incorrect. An infrared light will not cause the dye to glow.

    Answer C is correct. A blacklight (ultraviolet) will cause the dye to glow.

    Answer D is incorrect. A blue light will not cause the dye to glow.

28. Worn valve guides may cause all of these problems EXCEPT:

    A. Leaking of combustion gases.

    B. Excessive oil consumption.

    C. Uneven valve seating.

    D. Blowby.

TASK B.7

    Answer A is incorrect. Worn valve guides may cause improper valve seating, which can cause combustion gases to leak past the valve.

    Answer B is incorrect. Worn valve guides can cause oil consumption as the valve tips in the guide; excessive oil will run down the valve stem through the valve stem seal.

    Answer C is incorrect. Worn valve guides will allow uneven valve-to-seat contact causing the valve seat to wear and the valve to seat unevenly.

    Answer D is correct. Worn valve guides will not cause blowby; worn rings will cause this.

TASK C.9

29. Technician A says improper balance shaft timing can cause severe engine vibrations. Technician B says balance shafts are always timed in relation to the camshaft. Who is correct?

A.   A only

B.   B only

C.   Both A and B

D.   Neither A nor B

Answer A is correct. Only Technician A is correct. If the balance shaft is not properly timed, it will magnify normal engine vibrations.

Answer B is incorrect. Balance shafts are usually timed in relation to the crankshaft.

Answer C is incorrect. Only Technician A is correct.

Answer D is incorrect. Technician A is correct.

TASK B.9

30. The valve stem installed height is too tall. All of the following may be used to correct the installed height EXCEPT:

A.   Cutting the tip of the valve.

B.   Replacing the seat.

C.   Shimming the valve.

D.   Replacing the valve.

Answer A is incorrect. The tip of the valve may be cut to correct installed stem height.

Answer B is incorrect. As the seat is reconditioned, it causes the valve to seat deeper into it increasing stem installed height. If the seat has been cut too much, it may require replacement of the seat to correct the stem installed height.

Answer C is correct. The valve spring may need to be shimmed, but the valve cannot be shimmed.

Answer D is incorrect. Replacing the valve can be used to correct the stem installed height.

TASK E.7

31. All of the following would be a result of a plugged catalytic converter EXCEPT:

A.   Stalling.

B.   Loss of power.

C.   Power increase.

D.   Engine overheating.

Answer A is incorrect. Stalling may be a result of a plugged converter. Restriction to exhaust flow may cause exhaust gas to enter the intake manifold during valve overlap, causing stalling.

Answer B is incorrect. A partially plugged converter will not allow exhaust to be removed efficiently from the cylinder, causing a loss of power.

Answer C is correct. A plugged converter will cause a power decrease, or no power at all, since the cylinder cannot empty the burned gases.

Answer D is incorrect. A plugged converter can cause engine overheating. The exhaust removes about 30 percent of the heat produced during combustion; a plugged converter will prevent heat removal.

32. Technician A says an oil filter may contain an anti-drainback valve. Technician B says an oil filter will contain a bypass valve. Who is correct?

TASK D.4

   A.   A only
   B.   B only
   C.   Both A and B
   D.   Neither A nor B

Answer A is incorrect. Technician B is also correct.

Answer B is incorrect. Technician A is also correct.

Answer C is correct. Both Technicians are correct. The anti-drainback valve in an oil filter will not allow oil to run back to the oil pan when the engine is off. The bypass valve allows much of the oil to bypass the filtering material, under certain conditions, and go directly into the oil galleys.

Answer D is incorrect. Both Technicians are correct.

33. The customer complains of a knocking noise during cold starts in the morning that goes away within a minute of operation. Technician A says this could be caused by excessive connecting rod bearing clearance. Technician B says this could be piston slap caused by worn piston skirts. Who is correct?

TASK A.4

   A.   A only
   B.   B only
   C.   Both A and B
   D.   Neither A nor B

Answer A is incorrect. Connecting rod noise will not go away during warm-up and will increase in intensity as the oil warms up.

Answer B is correct. Only Technician B is correct. Piston slap is characterized by a knocking or slapping sound from the engine during cold starts that goes away as the engine warms up and the piston expands.

Answer C is incorrect. Only Technician B is correct.

Answer D is incorrect. Technician B is correct.

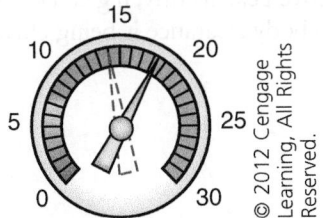

© 2012 Cengage Learning, All Rights Reserved.

34. With the engine idling, a vacuum gauge connected to the intake manifold fluctuates as shown in the figure above. These fluctuations may be caused by:

TASK A.6

   A.   Late ignition timing.
   B.   A restricted exhaust system.
   C.   Intake manifold leak at the throttle body.
   D.   Sticky valves.

Answer A is incorrect. Late ignition timing would result in a low, steady reading.

Answer B is incorrect. An intake manifold leak at the throttle body would cause a very low, steady reading.

Answer C is incorrect. A restricted exhaust system would cause vacuum to slowly decrease after engine RPM was raised and held steady.

Answer D is correct. If the valve sticks in its guide, it may not completely close resulting in a compression loss and pulsing vacuum gauge.

TASK C.12

35. Technician A says that when you are measuring ring end gap, it is only necessary to measure one of each ring set to verify correct fit. Technician B says to put each individual ring in the cylinder it will be in and measure ring end gap. Who is correct?

A.　A only

B.　B only

C.　Both A and B

D.　Neither A nor B

Answer A is incorrect. All rings should be measured for proper end gap.

Answer B is correct. Only Technician B is correct. All rings should be measured in the cylinder they will be in to ensure proper end gap and adjust for small differences between rings and cylinders.

Answer C is incorrect. Only Technician B is correct.

Answer D is incorrect. Technician B is correct.

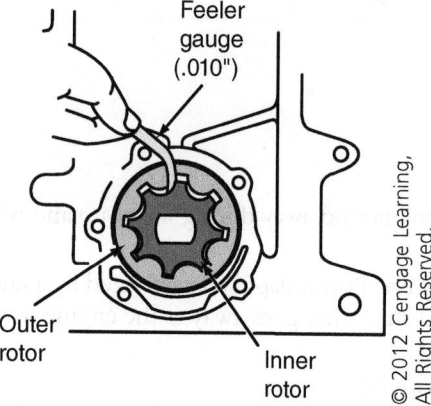

Feeler gauge (.010")

Outer rotor

Inner rotor

© 2012 Cengage Learning, All Rights Reserved.

TASK D.2

36. Technician A says that in the figure above, the oil pump drive gear-to-driven gear clearance is being checked. Technician B says the drive gear-to-pump body clearance is being checked. Who is correct?

A.　A only

B.　B only

C.　Both A and B

D.　Neither A nor B

Answer A is correct. Only Technician A is correct. The drive gear-to-driven gear clearance is being checked.

Answer B is incorrect. The drive-to-driven gear clearance is being checked.

Answer C is incorrect. Only Technician A is correct.

Answer D is incorrect. Technician A is correct.

37. Technician A says the valve seat must be refinished before any other head repairs are done. Technician B says the valve guide must be checked and repaired before the seat can be cut. Who is correct?

TASK B.8

   A.   A only
   B.   B only
   C.   Both A and B
   D.   Neither A nor B

   Answer A is incorrect. The valve guide must be repaired first.

   Answer B is correct. Only Technician B is correct. The valve guide forms the center point for the valve seat so it must be repaired before the valve seat can be properly repaired.

   Answer C is incorrect. Only Technician B is correct.

   Answer D is incorrect. Technician B is correct.

38. Technician A says a special puller and installer tool may be required to remove and install the harmonic balancer. Technician B says if there is deterioration of the rubber inertia ring, the balancer must be replaced. Who is correct?

TASK C.13

   A.   A only
   B.   B only
   C.   Both A and B
   D.   Neither A nor B

   Answer A is incorrect. Technician B is also correct.

   Answer B is incorrect. Technician A is also correct.

   Answer C is correct. Both Technicians are correct. Special pullers are necessary to remove and reinstall the balancer. Any deterioration of the inertia ring will cause failure of the balancer and increase vibrations.

   Answer D is incorrect. Both Technicians are correct.

39. A cylinder power balance test is being performed to determine which cylinder is causing a misfire. Technician A says when the misfiring cylinder is disabled, the RPM will drop little, or not at all. Technician B says the misfiring cylinder will cause an RPM increase when it is shorted. Who is correct?

TASK A.7

   A.   A only
   B.   B only
   C.   Both A and B
   D.   Neither A nor B

   Answer A is correct. Only Technician A is correct. During a power balance test, RPM drop is noted as each cylinder is shorted. If all cylinders are producing about the same power, the RPM drop should be within +/− 10 percent. If a cylinder is not producing adequate power, it is not contributing to the performance of the engine so there will be a much smaller, or no, drop in RPM.

   Answer B is incorrect. During a power balance test there will only be an RPM drop for each cylinder, no increase in RPM.

   Answer C is incorrect. Only Technician A is correct.

   Answer D is incorrect. Technician A is correct.

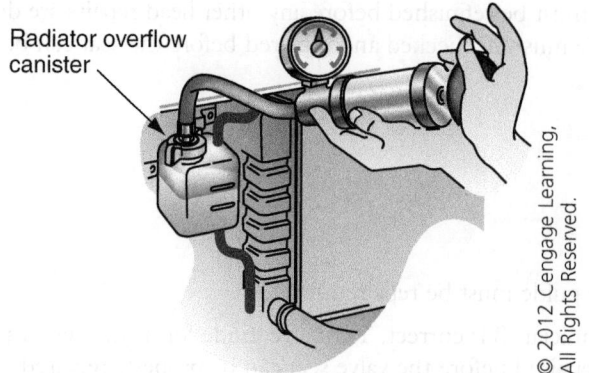

Radiator overflow canister

© 2012 Cengage Learning, All Rights Reserved.

TASK D.6

40. The tester in the figure above may be used to test all of the following EXCEPT:

    A.   Cooling system leaks.

    B.   Radiator cap pressure relief valve.

    C.   Heater core leaks.

    D.   Coolant specific gravity.

Answer A is incorrect. The coolant pressure tester is used to find coolant leaks.

Answer B is incorrect. The coolant pressure tester can be used to find a defective radiator cap pressure relief valve.

Answer C is incorrect. The coolant pressure tester can be used to find a leaking heater core.

Answer D is correct. The coolant pressure tester will not test specific gravity; a refractometer is used for this.

TASK C.10

41. Technician A says the proper place to measure piston diameter is at the crown above the first ring land. Technician B says the proper place to measure piston diameter is on the trunk, just below the piston pin. Who is correct?

    A.   A only

    B.   B only

    C.   Both A and B

    D.   Neither A nor B

Answer A is incorrect. The proper place to measure piston diameter is on the skirt.

Answer B is incorrect. The proper place to measure piston diameter is on the skirt.

Answer C is incorrect. Neither Technician is correct.

Answer D is correct. Neither Technician is correct. Piston diameter is measured on the trunk or skirt of the piston 90 degrees from the piston pin, close to the bottom.

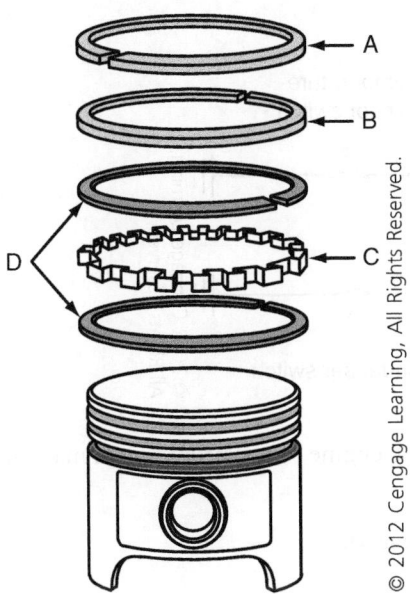

© 2012 Cengage Learning, All Rights Reserved.

42. In the figure above, Technician A says that C is part of the oil control ring set. Technician B says that A is the top compression ring and may have a mark on it showing the top of the ring. Who is correct?

TASK C.12

   A.   A only

   B.   B only

   C.   Both A and B

   D.   Neither A nor B

Answer A is incorrect. Technician B is also correct.

Answer B is incorrect. Technician A is also correct.

Answer C is correct. Both Technicians are correct. C is the expander for the oil control ring set, and A is the top compression ring and may have an identifying mark on it for the top side.

Answer D is incorrect. Both Technicians are correct.

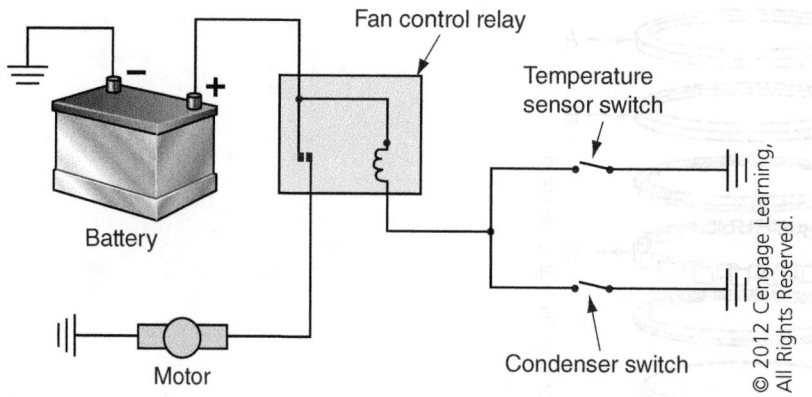

TASK D.12

43. In the figure above, an open ground circuit on the engine temperature switch may cause:

    A.   Continual cooling fan operation.

    B.   The fan not to operate when the A/C is engaged.

    C.   A burned out cooling fan motor.

    D.   Engine overheating.

Answer A is incorrect. An open in the circuit will cause the fan not to run.

Answer B is incorrect. The condenser switch will ground the fan circuit when the A/C is on.

Answer C is incorrect. An open ground circuit will not burn out the cooling fan motor.

Answer D is correct. If the cooling fan does not come on at the specified temperature, the engine may overheat.

TASK A.9

44. A cylinder leakage test has been performed and one cylinder failed with a leakage of over 50 percent. The technician notes that there are bubbles coming up in the radiator. What is the most likely source of the leakage?

    A.   A leaking intake valve

    B.   Worn rings

    C.   Cracked head

    D.   Leaking head gasket

Answer A is incorrect. A leaking intake valve would allow air to escape past the throttle plate.

Answer B is incorrect. Worn rings will allow air to go to the oil pan and escaping air would be heard at the oil dipstick.

Answer C is incorrect. Although a cracked head could cause leakage into the cooling system, it is not the most likely source.

Answer D is correct. A head gasket that has failed between the cylinder and a water passage will allow air to be pushed into the cooling system; the air will show up as bubbles in the radiator.

45. A small pickup with a 4-cylinder engine is brought into the shop for a rattling noise from the front of the engine. Technician A says the noise could be caused by a timing chain that has too much slack. Technician B says the timing chain guides could be worn out. Who is correct?

TASK A.4

A. A only

B. B only

C. Both A and B

D. Neither A nor B

Answer A is incorrect. Technician B is also correct.

Answer B is incorrect. Technician A is also correct.

Answer C is correct. Both Technicians are correct. A stretched timing chain will cause a rattling noise at the front of the engine. The timing chain guides may also be worn out causing the chain to rattle.

Answer D is incorrect. Both Technicians are correct.

46. Technician A says that when performing a cylinder leakage test, 10 percent leakage is acceptable. Technician B says the acceptable leakage is due to incomplete sealing of the rings. Who is correct?

TASK A.9

A. A only

B. B only

C. Both A and B

D. Neither A nor B

Answer A is incorrect. Technician B is also correct.

Answer B is incorrect. Technician A is also correct.

Answer C is correct. Both Technicians are correct. Piston rings have a specified installed ring gap. That gap would allow for 10 percent of leakage into the crankcase during a cylinder leakage test.

Answer D is incorrect. Both Technicians are correct.

47. Technician A says that a cylinder with 0.020 inch ring ridge can be honed and reused. Technician B says the ring ridge may have to be removed to remove the piston assembly. Who is correct?

TASK C.4

A. A only

B. B only

C. Both A and B

D. Neither A nor B

Answer A is incorrect. A cylinder with 0.020 inch ring ridge must be bored oversize.

Answer B is correct. Only Technician B is correct. Removing the piston without removing the ring ridge may cause ring or piston land breakage.

Answer C is incorrect. Only Technician B is correct.

Answer D is incorrect. Technician B is correct.

TASK A.4

48. An engine has a deep knock while running that increases in speed along with engine RPM. The most likely cause of the noise is:

   A. Worn lifters.

   B. A cracked flexplate.

   C. Water pump bearings.

   D. A connecting rod with too much bearing clearance.

   Answer A is incorrect. Worn lifter noise will be a light tapping noise in the top of the engine.

   Answer B is incorrect. A cracked flexplate will usually quiet down as engine RPM is increased and transmission pressure evens out.

   Answer C is incorrect. Worn water pump bearings will growl or grind, not knock.

   Answer D is correct. A connecting rod with excessive bearing clearance will cause a deep knocking noise from the bottom of the engine that will increase with engine RPM.

TASK D.6

49. Technician A says a defective radiator cap pressure valve may cause an engine to run too cold. Technician B says a defective radiator cap vacuum valve may cause the upper radiator hose to collapse. Who is correct?

   A. A only

   B. B only

   C. Both A and B

   D. Neither A nor B

   Answer A is incorrect. A defective pressure valve may cause the coolant to boil over when the engine is turned off. The resulting loss of coolant could cause the engine to overheat.

   Answer B is correct. Only Technician B is correct. A defective vacuum valve that does not open may cause the upper hose to collapse. As the coolant cools and contracts, a vacuum can be formed in the cooling system.

   Answer C is incorrect. Only Technician B is correct.

   Answer D is incorrect. Technician B is correct.

TASK A.2

50. The vehicle will not crank. Technician A says this may be caused by hydrolock in a cylinder. Technician B says the fuel pump may be defective. Who is correct?

   A. A only

   B. B only

   C. Both A and B

   D. Neither A nor B

   Answer A is correct. Only Technician A is correct. Hydrolock in a cylinder will not allow the engine to spin over.

   Answer B is incorrect. The fuel pump cannot cause an engine not to crank. It will cause an engine not to start.

   Answer C is incorrect. Only Technician A is correct.

   Answer D is incorrect. Technician A is correct.

PREPARATION EXAM 2—ANSWER KEY

1.	C	21.	C	41.	C
2.	A	22.	C	42.	D
3.	C	23.	A	43.	B
4.	A	24.	C	44.	C
5.	A	25.	A	45.	B
6.	B	26.	C	46.	C
7.	B	27.	C	47.	D
8.	D	28.	C	48.	C
9.	D	29.	B	49.	B
10.	C	30.	C	50.	C
11.	B	31.	C		
12.	C	32.	D		
13.	C	33.	C		
14.	C	34.	C		
15.	A	35.	D		
16.	B	36.	A		
17.	B	37.	B		
18.	B	38.	D		
19.	C	39.	A		
20.	C	40.	C		

PREPARATION EXAM 2—EXPLANATIONS

1. The customer's complaint is blue smoke emitting from the tailpipe on initial start up in the morning, which seems to dissipate quickly. Technician A says that the blue smoke is caused by oil burning in the combustion chamber. Technician B says the oil is probably entering the combustion chamber through deteriorated valve stem seals. Who is correct?

   TASK A.5

   A. A only
   B. B only
   C. Both A and B
   D. Neither A nor B

   Answer A is incorrect. Technician B is also correct.

   Answer B is incorrect. Technician A is also correct.

   Answer C is correct. Both Technicians are correct. Blue smoke that quickly dissipates on initial cold start is a sign that the valve stem seals are allowing oil to run down the valve stem into the combustion chamber after warm shutdown.

   Answer D is incorrect. Both technicians are correct.

TASK A.8

2. While performing a cranking compression test on a 4-cylinder engine, the technician notes that one cylinder has a pressure reading of 60 psi, 44 kPa, while the others have a reading of 135 psi, 931 kPa. Technician A says performing a cylinder leakage test will indicate where the pressure is leaking. Technician B says the vehicle has leaking valve stem seals. Who is correct?

A. A only

B. B only

C. Both A and B

D. Neither A nor B

Answer A is correct. Only Technician A is correct. The next step for narrowing down a cause for low compression on one cylinder would be the cylinder leakage test.

Answer B is incorrect. Leaking valve stem seals will not cause a loss of compression.

Answer C is incorrect. Only Technician A is correct.

Answer D is incorrect. Technician A is correct.

TASK A.5

3. The customer's complaint is excessive oil use and spots of oil on his driveway. Upon inspection, it appears that the valve cover gaskets, oil pan, and front main seal are leaking. Technician A says the problem is gasket deterioration, and they must be replaced. Technician B says a defective PCV valve may be causing the leaks. Who is correct?

A. A only

B. B only

C. Both A and B

D. Neither A nor B

Answer A is incorrect. Technician B is also correct.

Answer B is incorrect. Technician A is also correct.

Answer C is correct. Both Technicians are correct. The gaskets may be deteriorated and need replacement. A defective PVC valve will cause excessive crankcase pressures causing the seals and gaskets to leak.

Answer D is incorrect. Both Technicians are correct.

TASK D.1

4. The customer says his oil pressure gauge stays low even at highway speeds. Technician A says this may be caused by worn crankshaft main bearings. Technician B says this can be caused by leaking rings. Who is correct?

A. A only

B. B only

C. Both A and B

D. Neither A nor B

Answer A is correct. Only Technician A is correct. Worn main bearings will cause oil pressure to be low at all speeds.

Answer B is incorrect. Leaking rings will cause compression loss but will not affect oil pressure.

Answer C is incorrect. Only Technician A is correct.

Answer D is incorrect. Technician A is correct.

5. The customer's complaint is that the engine cranks over slowly but does not start. Technician A says it could be caused by the battery. Technician B says it could be caused by a broken timing belt. Who is correct?

    TASK A.2

    A. A only
    B. B only
    C. Both A and B
    D. Neither A nor B

    Answer A is correct. Only Technician A is correct. A weak battery would cause slow cranking.

    Answer B is incorrect. A broken timing belt will cause the engine to spin over rapidly.

    Answer C is incorrect. Only Technician A is correct.

    Answer D is incorrect. Technician A is correct.

6. Technician A says RTV can be used to seal lower intake plenum in place of a gasket. Technician B says RTV can be used to replace a water pump gasket. Who is correct?

    TASK C.16

    A. A only
    B. B only
    C. Both A and B
    D. Neither A nor B

    Answer A is incorrect. RTV will deteriorate when in contact with gasoline.

    Answer B is correct. Only Technician B is correct. RTV is a suitable replacement for a gasket on a water pump.

    Answer C is incorrect. Only Technician B is correct.

    Answer D is incorrect. Technician B is correct.

7. The customer complains of a vibration when accelerating, which goes away at a steady speed. Which of the following is the most likely cause for the condition?

    TASK A.4

    A. Out of balance tire
    B. Broken or weak motor mount
    C. Bent or broken engine cooling fan
    D. A warped brake rotor

    Answer A is incorrect. A vibration from an out of balance tire would be felt at speeds above 35 MPH rather than just when accelerating.

    Answer B is correct. A broken or weak motor mount will allow the engine to shift under acceleration, causing a vibration.

    Answer C is incorrect. A bent or broken cooling fan would cause a constant vibration.

    Answer D is incorrect. A warped brake rotor will cause a vibration when stopping, not when starting.

TASK C.1

8. When preparing an engine for removal, which of the following would be done?

    A.   Drain engine coolant.

    B.   Drain engine oil.

    C.   Disconnect fuel lines.

    D.   All of the above

    Answer A is incorrect. Because hoses are connected to the engine, draining the engine coolant is necessary.

    Answer B is incorrect. Because the oil pan and filter must be removed after removing the engine, the oil must be drained.

    Answer C is incorrect. Because fuel lines are connected to the engine, they must be disconnected before the engine can be removed.

    Answer D is correct. All the fluids should be removed from the engine and fuel lines disconnected.

TASK A.9

9. A technician is performing a cylinder leakage test and cylinder #3 has 45 percent leakage. Which is the LEAST LIKELY place the technician would look for escaping air?

    A.   The cooling system

    B.   The intake system

    C.   The exhaust system

    D.   The transmission dipstick

    Answer A is incorrect. The cooling system should be checked.

    Answer B is incorrect. The intake system should be checked.

    Answer C is incorrect. The exhaust system should be checked.

    Answer D is correct. The transmission will not show air leakage from a cylinder in the engine.

TASK E.5

10. Technician A says oil accumulation in the air cleaner could be caused by worn rings. Technician B says oil accumulation in the air cleaner could be caused by a defective PCV valve or hose. Who is correct?

    A.   A only

    B.   B only

    C.   Both A and B

    D.   Neither A nor B

    Answer A is incorrect. Technician B is also correct.

    Answer B is incorrect. Technician A is also correct.

    Answer C is correct. Both Technicians are correct. Worn rings can cause excessive blowby creating pressure in the crankcase and allowing oil to be blown into the air cleaner. A defective PCV system will do much the same thing.

    Answer D is incorrect. Both Technicians are correct.

11. A cylinder power balance test is being performed. Technician A says a compression gauge is used for a power balance test. Technician B says a misfiring cylinder will cause a small RPM decrease or no change at all when it is shorted. Who is correct?

TASK A.7

A. A only

B. B only

C. Both A and B

D. Neither A nor B

Answer A is incorrect. A compression gauge is used for checking compression pressure.

Answer B is correct. Only Technician B is correct. During a power balance test, RPM drop is noted as each cylinder is shorted. If all cylinders are producing about the same power, the RPM drop should be within +/−10 percent. If a cylinder is not producing adequate power, it is not contributing to the performance of the engine so there will be a much smaller or no drop in RPM.

Answer C is incorrect. Only Technician B is correct.

Answer D is incorrect. Technician B is correct.

12. Technician A says a rod knock will become quieter as its cylinder is grounded. Technician B says a loose piston pin will cause a double click noise. Who is correct?

TASK A.4

A. A only

B. B only

C. Both A and B

D. Neither A nor B

Answer A is incorrect. Technician B is also correct.

Answer B is incorrect. Technician A is also correct.

Answer C is correct. Both Technicians are correct. When the plug wire to a cylinder is grounded, a rod knock will become quieter since it is not subject to the pressure of combustion. A loose piston pin will cause a double click as the piston reaches TDC or BDC and changes direction.

Answer D is incorrect. Both Technicians are correct.

13. A customer says his car will not go over 50 mph. The technician installs a vacuum gauge on the engine and notes a normal, steady vacuum at idle. When he brings the RPM to 2,000 rpm and holds it there, he notes a steady drop in vacuum. Which of the following is the most likely cause?

TASK E.7

A. A leaking intake manifold

B. A misfiring spark plug

C. A plugged exhaust system

D. Weak rings

Answer A is incorrect. A leaking intake manifold will cause low vacuum at idle.

Answer B is incorrect. A misfiring spark plug would cause a rough idle and a fluctuating vacuum.

Answer C is correct. A plugged exhaust system will cause excessive back-pressure and the exhaust gases will not be efficiently removed from the cylinder, diluting the intake mixture and causing poor power production.

Answer D is incorrect. Weak rings will cause a low vacuum reading at idle.

TASK D.5

14. There is coolant leaking from an engine compartment but the technician cannot tell where it is coming from. Technician A says dye could be put into the cooling system and the leak will show up under inspection with a blacklight. Technician B says to pressurize the cooling system then look for leaks. Who is correct?

A. A only

B. B only

C. Both A and B

D. Neither A nor B

Answer A is incorrect. Technician B is also correct.

Answer B is incorrect. Technician A is also correct.

Answer C is correct. Both Technicians are correct. A special dye put into the cooling system will circulate with the coolant, and the point of leak will be evident under a blacklight. A coolant pressure test is used to find coolant leaks. It can hold pressure on the system while the technician looks for the leak.

Answer D is incorrect. Both Technicians are correct.

TASK C.7

15. Technician A says main bearing oil clearance can be checked with Plastigauge®. Technician B says main bearing oil clearance can be checked with a feeler gauge. Who is correct?

A. A only

B. B only

C. Both A and B

D. Neither A nor B

Answer A is correct. Only Technician A is correct. Main bearing oil clearance can be checked with Plastigauge® by placing the Plastigauge® across the journal with the bearings installed and torqueing the cap; remove the cap, and use the Plastigauge® scale, record the clearance.

Answer B is incorrect. A feeler gauge is not a tool to check main bearing oil clearance. It will not conform to the journal to give an accurate reading.

Answer C is incorrect. Only Technician A is correct.

Answer D is incorrect. Technician A is correct.

TASK E.6

16. The spark plugs are being replaced. Technician A says platinum spark plug gap does not have to be checked. Technician B says a replacement spark plug must be the same heat range and style as the original. Who is correct?

A. A only

B. B only

C. Both A and B

D. Neither A nor B

Answer A is incorrect. Spark plug gap should always be carefully checked to ensure the gap is correct. Although platinum plugs are pre-gapped at the factory, the plug may have been dropped, causing the gap to close.

Answer B is correct. Only Technician B is correct. Although there are many different manufacturers for spark plugs, the replacement plugs must be the same heat range and composition as the originals to provide good spark for extended time without failure.

Answer C is incorrect. Only Technician B is correct.

Answer D is incorrect. Technician B is correct.

17. Florescent dye has been added to the crankcase to help locate an oil leak. Technician A says the dye will be visible with an infrared light at the point of the leak. Technician B says a blacklight will cause the dye to be visible. Who is correct?

TASK A.3

   A.   A only
   B.   B only
   C.   Both A and B
   D.   Neither A nor b

   Answer A is incorrect. An infrared light will not cause the dye to glow.

   Answer B is correct. Only Technician B is correct. A blacklight will cause the florescent dye to glow.

   Answer C is incorrect. Only Technician B is correct.

   Answer D is incorrect. Technician B is correct.

18. Technician A says a light coat of RTV should be applied to a rubber valve cover gasket during installation to ensure a good seal. Technician B says RTV is a gasket maker and should not be applied to the rubber valve cover gasket. Who is correct?

TASK C.15

   A.   A only
   B.   B only
   C.   Both A and B
   D.   Neither A nor B

   Answer A is incorrect. RTV is a gasket material. A rubber valve cover gasket and applying RTV means a double gasket. RTV also can make a rubber gasket slip out of position because RTV is slippery until it is cured.

   Answer B is correct. Only Technician B is correct. RTV is made to be used by itself as a gasket.

   Answer C is incorrect. Only Technician B is correct.

   Answer D is incorrect. Technician B is correct.

19. The threads in a camshaft tower hole have been damaged. Technician A says the threads may be tapped with the proper tap size and thread pitch to restore the threads. Technician B says the threads may have to be restored using a HeliCoil thread insert. Who is correct?

TASK B.3

   A.   A only
   B.   B only
   C.   Both A and B
   D.   Neither A nor B

   Answer A is incorrect. Technician B is also correct.

   Answer B is incorrect. Technician A is also correct.

   Answer C is correct. Both Technicians are correct. If the damage is not too bad, the hole may be tapped with the correct tap to refresh the threads. If there is not enough metal left for rethreading, a HeliCoil thread insert can be used.

   Answer D is incorrect. Both Technicians are correct.

TASK C.15

20. An oil pan is being reinstalled on the engine. Technician A says that the oil pan can be sealed using RTV. Technician B says that a new gasket must be installed. Who is correct?

A. A only

B. B only

C. Both A and B

D. Neither A nor B

Answer A is incorrect. Technician B is also correct.

Answer B is incorrect. Technician A is also correct.

Answer C is correct. Both Technicians are correct. RTV is a gasket maker so RTV or a new gasket could be installed.

Answer D is incorrect. Both Technicians are correct.

TASK C.9

21. Technician A says balance shafts rotate in the opposite direction of the crankshaft. Technician B says the balance shafts are always timed in relation to the crankshaft. Who is correct?

A. A only

B. B only

C. Both A and B

D. Neither A nor B

Answer A is incorrect. Technician B is also correct.

Answer B is incorrect. Technician A is also correct.

Answer C is correct. Both Technicians are correct. Balance shafts rotate in the opposite direction of the crankshaft. Balance shafts are always timed in relation to the crankshaft.

Answer D is incorrect. Both Technicians are correct.

TASK B.13

22. The customer is concerned about poor performance at highway speeds. The technician talks to the owner about the car's repair history and is told that the timing belt, water pump, and crank seal have recently been replaced, and the problem started afterward. What is the most likely cause for poor performance?

A. The serpentine belt is loose.

B. The sparkplug wires were misrouted.

C. The camshaft and crankshaft are out of sync.

D. The catalytic converter is restricted.

Answer A is incorrect. A loose serpentine belt will cause a squeal but not affect performance.

Answer B is incorrect. Misrouted sparkplug wires would cause a misfire and poor performance all the time.

Answer C is correct. If the camshaft and crankshaft were not positioned properly during timing belt replacement, performance could suffer at higher engine RPM.

Answer D is incorrect. A restricted converter would case performance problems at any speed.

23. During disassembly of the block, the technician notes that main bearing wear is greater on the first journal's top bearing and the bottom bearing of the last journal. Technician A says this may have been caused by accessory belts being overtightened. Technician B says this was caused by oil starvation. Who is correct?

    TASK C.5

    A. A only
    B. B only
    C. Both A and B
    D. Neither A nor B

    Answer A is correct. Only Technician A is correct. If an accessory belt is overtightened, it will pull up on the front of the crankshaft, causing upper bearing wear on the first main bearing journal, and lower bearing wear on the last main journal.

    Answer B is incorrect. Oil starvation would be evident on the lower half of all the main bearings.

    Answer C is incorrect. Only Technician A is correct.

    Answer D is incorrect. Technician A is correct.

24. Technician A says that when the fuel lines are reconnected on a replaced engine, the o-ring seals on the fuel lines must be replaced. Technician B says the injector o-rings should be replaced when they are reinstalled. Who is correct?

    TASK E.1

    A. A only
    B. B only
    C. Both A and B
    D. Neither A nor B

    Answer A is incorrect. Technician B is also correct.

    Answer B is incorrect. Technician A is also correct.

    Answer C is correct. Both Technicians are correct. Both the fuel line seals and injector seals must be replaced to ensure there will be no fuel leakage.

    Answer D is incorrect. Both Technicians are correct.

25. All of the following could prevent a starter solenoid from engaging EXCEPT:

    TASK E.4

    A. A bent armature shaft.
    B. A shorted clutch safety switch.
    C. An open ignition switch.
    D. An open in the neutral safety switch.

    Answer A is correct. A bent armature shaft would not keep the starter solenoid from engaging, although it may prevent the starter from turning.

    Answer B is incorrect. A shorted clutch safety switch may ground the power to the starter relay causing the solenoid to not engage.

    Answer C is incorrect. An open ignition switch will not provide power to the starter circuit.

    Answer D is incorrect. An open in the neutral safety switch will not allow power to flow to the starter relay.

TASK C.11

26. Technician A says during disassembly of the engine block, all parts must be kept in order for later inspection. Technician B says connecting rod caps should be marked to identify which cylinder they came out of. Who is correct?

    A.   A only

    B.   B only

    C.   Both A and B

    D.   Neither A nor B

Answer A is incorrect. Technician B is also correct.

Answer B is incorrect. Technician A is also correct.

Answer C is correct. Both Technicians are correct. During teardown, all parts should be kept in order for inspection and fault diagnosis. If the connecting rods on a V-engine are put on the opposite side of the block they came from, piston drag and binding may occur.

Answer D is incorrect. Both Technicians are correct.

TASK A.6

27. Technician A says a vacuum gauge reading of 16 to 21 in. Hg at idle is normal. Technician B says a vacuum gauge reading that is steady but low could indicate retarded valve timing. Who is correct?

    A.   A only

    B.   B only

    C.   Both A and B

    D.   Neither A nor B

Answer A is incorrect. Technician B is also correct.

Answer B is incorrect. Technician A is also correct.

Answer C is correct. Both Technicians are correct. Normal vacuum reading will range from 16 to 21 in. Hg depending on the size and configuration of the engine. A steady but low vacuum reading could be caused by valves opening late due to retarded camshaft-to-crankshaft relationship.

Answer D is incorrect. Both Technicians are correct.

TASK B.9

28. Technician A says too tall of an installed valve stem height could cause burned valves. Technician B says too short of an installed valve stem height could cause poor performance. Who is correct?

    A.   A only

    B.   B only

    C.   Both A and B

    D.   Neither A nor B

Answer A is incorrect. Technician B is also correct.

Answer B is incorrect. Technician A is also correct.

Answer C is correct. Both Technicians are correct. Too tall of an installed height may not allow the valve to close fully, causing the valve to burn. Too short of an installed height will not allow the valve to be open long enough to properly fill the cylinder.

Answer D is incorrect. Both Technicians are correct.

29. Technician A says that cylinder walls that do not require boring can be cleaned and reused without further servicing. Technician B says the cylinder walls should be deglazed and cleaned to provide oil retention during break-in. Who is correct?

    A. A only
    B. B only
    C. Both A and B
    D. Neither A nor B

    TASK C.4

    Answer A is incorrect. The cylinder walls should be deglazed to properly seat new rings.
    Answer B is correct. Only Technician B is correct. The cylinder walls should be deglazed and cleaned to help the new rings seat properly.
    Answer C is incorrect. Only Technician B is correct.
    Answer D is incorrect. Technician B is correct.

30. Technician A says talking to the customer to make sure you understand his complaint is a good practice. Technician B says it is important to test drive the vehicle to duplicate the customer's complaint. Who is correct?

    A. A only
    B. B only
    C. Both A and B
    D. Neither A nor B

    TASK A.1

    Answer A is incorrect. Technician B is also correct.
    Answer B is incorrect. Technician A is also correct.
    Answer C is correct. Both Technicians are correct. A clear understanding of the customer's problem will save time in identifying the problem, and a test drive is often necessary to verify and understand the complaint.
    Answer D is incorrect. Both Technicians are correct.

© 2012 Cengage Learning, All Rights Reserved.

31. In the figure above, the technician is most likely checking:

    A. Valve guide wear.
    B. Valve stem installed height.
    C. Valve seat concentricity.
    D. Valve seat angle.

    TASK B.8

    Answer A is incorrect. Valve guide wear would be checked with a small hole gauge and dial caliper.
    Answer B is incorrect. Valve stem installed height is checked with the valve installed in the seat and the distance between the head surface or spring seat and the valve tip is measured.
    Answer C is correct. The technician is checking valve seat concentricity with a concentricity gauge installed on a pilot in the valve guide.
    Answer D is incorrect. Valve seat angle is set by using the proper angle seat resurfacing tool.

TASK D.6

32. An upper radiator hose that collapses as the engine cools down can be caused by:

    A. A defective radiator cap pressure seal.

    B. A faulty upper radiator hose.

    C. A stuck open thermostat.

    D. A defective radiator cap vacuum valve.

Answer A is incorrect. A defective radiator cap pressure seal can cause coolant to boil over.

Answer B is incorrect. A faulty radiator hose will swell under pressure.

Answer C is incorrect. A stuck open thermostat will not allow the engine to reach operating temperature as soon as it should.

Answer D is correct. A defective radiator cap vacuum valve will cause a vacuum to be developed in the cooling system when it cools down and may cause the upper hose to collapse.

TASK C.6

33. Technician A says during teardown, the main and rod bearings should be kept in order to identify abnormal wear. Technician B says the main bearing bore should be checked for misalignment. Who is correct?

    A. A only

    B. B only

    C. Both A and B

    D. Neither A nor B

Answer A is incorrect. Technician B is also correct.

Answer B is incorrect. Technician A is also correct.

Answer C is correct. Both Technicians are correct. Keeping the main and rod bearings in order allows the technician to identify abnormal wear patterns. If the wear pattern indicates possible misalignment in the main bore, it should be checked with a straightedge and feeler gauge.

Answer D is incorrect. Both Technicians are correct.

TASK A.5

34. An excessive sulfur smell in the exhaust of a vehicle with a catalytic converter can be an indication of:

    A. A lean fuel mixture.

    B. Coolant leaking into the combustion chamber.

    C. A rich fuel mixture.

    D. A vacuum leak.

Answer A is incorrect. A lean fuel mixture would not cause a sulfur smell.

Answer B is incorrect. Coolant leaking into the combustion chamber would cause white exhaust color with a sweet smell.

Answer C is correct. A rich fuel mixture will cause a strong sulfur smell from the exhaust as the extra fuel is burned in the converter.

Answer D is incorrect. A vacuum leak would cause a rough idle that would decrease as engine speed increases.

35. Technician A says that when heads are removed from the block, it does not matter how the head bolts are removed. Technician B says it is best to remove the heads while they are still warm to prevent warping on the deck surface. Who is correct?

A. A only

B. B only

C. Both A and B

D. Neither A nor B

TASK B.1

Answer A is incorrect. Head bolts should be removed in the reverse order of the torque sequence.

Answer B is incorrect. Removing a head while it is still warm may cause deck surface warpage.

Answer C is incorrect. Neither Technician is correct.

Answer D is correct. Neither Technician is correct. Head bolts should be removed in the reverse order of the torque sequence; removing a head while it is still warm may cause it to warp.

36. Technician A says that some serpentine belts can be easily misrouted causing belt squeal. Technician B says most vehicles have a belt routing diagram in the owner's manual. Who is correct?

A. A only

B. B only

C. Both A and B

D. Neither A nor B

TASK D.7

Answer A is correct. Only Technician A is correct. Serpentine belt routing can be complicated because of the route it must take to maintain sufficient tension on all the accessories to turn them under all loads.

Answer B is incorrect. Most manufacturers put a placard under the hood with the belt routing diagram. There is no diagram in the owner's manual.

Answer C is incorrect. Only Technician A is correct.

Answer D is incorrect. Technician A is correct.

37. When doing a compression test, the results for all cylinders are all even, but lower than the specified compression pressure. This could indicate:

A. A blown head gasket.

B. Worn rings and cylinders.

C. A cracked head.

D. Carbon buildup on the pistons.

TASK A.8

Answer A is incorrect. A blown head gasket would not typically affect all cylinders.

Answer B is correct. If worn rings and cylinders are suspected, do a wet compression test on all cylinders. If the compression pressure increases, this is a good indication of worn rings.

Answer C is incorrect. A cracked head will not affect all cylinders.

Answer D is incorrect. Carbon buildup on the pistons will cause increased compression.

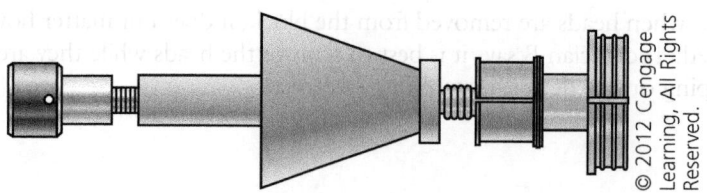

© 2012 Cengage Learning, All Rights Reserved.

TASK B.15

38. The tool shown in the figure above is used to:

  A. Remove crankshaft bearings.

  B. Remove camshaft bearings only.

  C. Remove pistons.

  D. Remove and install camshaft bearings.

  Answer A is incorrect. This tool is used for camshaft bearing removal and installation.

  Answer B is incorrect. The tool is also used to install camshaft bearings.

  Answer C is incorrect. This tool is used for camshaft bearing removal and installation.

  Answer D is correct. This tool is used for both removal and installation of camshaft bearings.

TASK A.5

39. A vehicle is brought in with excessive oil consumption. Technician A says a lack of regular oil changes can cause the oil rings to stick. Technician B says a burned valve can cause excessive oil consumption. Who is correct?

  A. A only

  B. B only

  C. Both A and B

  D. Neither A nor B

  Answer A is correct. Only Technician A is correct. Lack of regular oil changes will cause varnish buildup on the internal engine components. This varnish buildup can cause oil control rings to stick in their groove and leave oil on the cylinder wall, which is burned during combustion.

  Answer B is incorrect. A burned valve will cause poor engine performance, but will not affect excessive oil consumption.

  Answer C is incorrect. Only Technician A is correct.

  Answer D is incorrect. Technician A is correct.

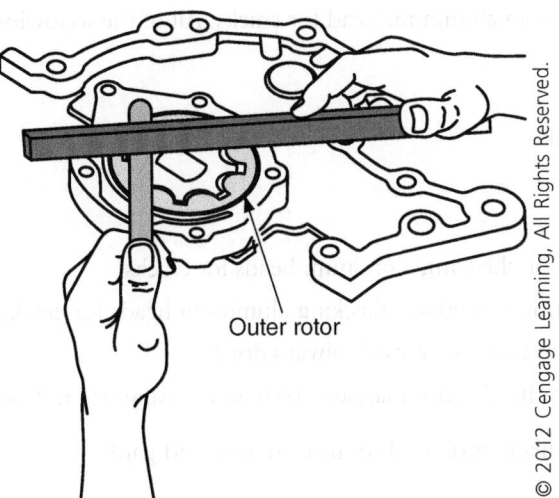

Outer rotor

© 2012 Cengage Learning, All Rights Reserved.

40. The technician in the figure above is most likely:

    A. Checking oil pump drive gear-to-driven gear clearance.

    B. Checking oil pump housing trueness.

    C. Checking oil pump gear-to-body clearance.

    D. Checking oil pump pressure relief valve.

TASK D.2

Answer A is incorrect. Oil pump drive gear-to-driven gear clearance is typically checked with a gauge rod placed between the two oil pump gears.

Answer B is incorrect. Checking the oil pump housing trueness would be done between the straightedge and the housing surfaces, not the gear assembly.

Answer C is correct. The technician is checking for excessive clearance between the pump gears and the body of the pump.

Answer D is incorrect. The oil pump pressure relief valve is mounted on the outside of the pump housing and would be checked by testing spring pressure.

41. Technician A says that most engine cooling fans are controlled through the ECM. Technician B says the cooling fan can be turned on based on engine temperature or A/C selection. Who is correct?

    A. A only

    B. B only

    C. Both A and B

    D. Neither A nor B

TASK D.12

Answer A is incorrect. Technician B is also correct.

Answer B is incorrect. Technician A is also correct.

Answer C is correct. Both Technicians are correct. Modern cooling fans are enabled through the ECM based on engine temperature or A/C selection.

Answer D is incorrect. Both Technicians are correct.

TASK B.2

42. The technician is preparing to check an aluminum head for cracks. All of the following are acceptable methods EXCEPT:

A. Dye.

B. Pressure testing in a water tank.

C. Visual inspection.

D. Magnafluxing.

Answer A is incorrect. Dye is used for checking aluminum heads for cracks.

Answer B is incorrect. Pressure testing is used for checking aluminum heads for cracks.

Answer C is incorrect. A visual inspection for cracks is always done.

Answer D is correct. An aluminum head is not magnetic so magnafluxing will not work.

TASK C.12

43. Which of these would be the most likely tool used to measure ring end gap?

A. Outside micrometer

B. Feeler gauge

C. Dial caliper

D. Dial indicator

Answer A is incorrect. An outside micrometer is used to measure outside diameters.

Answer B is correct. Ring end gap is measured with a feeler gauge.

Answer C is incorrect. A dial caliper is used for outside, inside, or depth measurements.

Answer D is incorrect. A dial indicator is used to measure end-play.

TASK D.9

44. If a thermostat fails in the open position, all of the following could occur EXCEPT:

A. Erratic computer control system operation.

B. Poor fuel economy.

C. Loss of coolant.

D. Longer than normal warmup period.

Answer A is incorrect. An open thermostat will cause the engine to be slow, or to never reach normal operating temperature, causing the computer to stay in open loop longer.

Answer B is incorrect. An open thermostat causing slow engine warm-up will cause poor fuel economy.

Answer C is correct. An open thermostat will not cause boil over and loss of coolant.

Answer D is incorrect. Longer than normal warm-up is a result of a thermostat failing in the open position.

45. Technician A says that valve train related noise is heard as a heavy knocking noise in the middle of the engine. Technician B says that valve train related noise is a light tapping or rattling noise in the top of the engine. Who is correct?

    TASK A.4

    A. A only
    B. B only
    C. Both A and B
    D. Neither A nor B

    Answer A is incorrect. Valve train noise comes from the top of the engine and is heard as a light tapping or rattling noise.

    Answer B is correct. Only Technician B is correct. Valve train noise is a light rattling or tapping noise from the top of the engine.

    Answer C is incorrect. Only Technician B is correct.

    Answer D is incorrect. Technician B is correct.

46. Technician A says balance shafts may be mounted above the camshaft. Technician B says the balance shafts may be mounted on the bottom of the engine. Who is correct?

    TASK C.9

    A. A only
    B. B only
    C. Both A and B
    D. Neither A nor B

    Answer A is incorrect. Technician B is also correct.

    Answer B is incorrect. Technician A is also correct.

    Answer C is correct. Both Technicians are correct. Balance shafts can be mounted above the camshaft or on the bottom of the engine, depending on engine design.

    Answer D is incorrect. Both Technicians are correct.

47. Technician A says an ethylene glycol and water mixture raises the freezing point of the coolant. Technician B says an ethylene glycol and water mixture lowers the boiling point of the coolant. Who is correct?

    TASK D.10

    A. A only
    B. B only
    C. Both A and B
    D. Neither A nor B

    Answer A is incorrect. The proper mixture of ethylene glycol and water lowers the freezing point of the coolant.

    Answer B is incorrect. The proper mixture of ethylene glycol and water raises the boiling point of the coolant.

    Answer C is incorrect. Neither Technician is correct.

    Answer D is correct. Neither Technician is correct. The proper mixture lowers the freezing point and raises the boiling point.

TASK B.2

48. Upon head removal, the technician notes the coolant passage holes in the head gasket have corroded out to the same size as the block and head passages. Technician A says this could cause engine overheating at high speeds. Technician B says the holes in the gasket are substantially smaller than the coolant passage holes in the block and head. Who is correct?

   A. A only

   B. B only

   C. Both A and B

   D. Neither A nor B

Answer A is incorrect. Technician B is also correct.

Answer B is incorrect. Technician A is also correct.

Answer C is correct. Both Technicians are correct. Enlarged coolant passage holes in the head gasket may allow coolant, at high RPM, to pass through the head so fast that it does not pick up enough heat to cause the engine to overheat. The holes in the head gasket are made substantially smaller to slow coolant flow for better heat transfer to prevent overheating.

Answer D is incorrect. Both Technicians are correct.

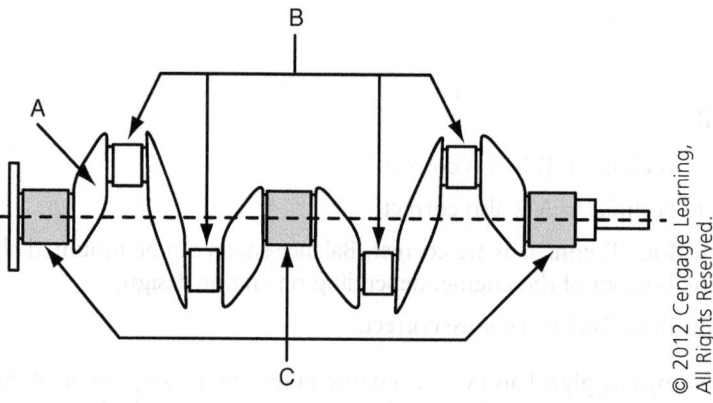

© 2012 Cengage Learning, All Rights Reserved.

TASK C.5

49. Referring to the crankshaft in the figure above, which of the statements below is the LEAST LIKELY:

   A. B indicates the rod journals.

   B. C indicates the crankshaft thrust surface.

   C. A indicates cross-drilled oil supply holes.

   D. C indicates the main bearing journals.

Answer A is incorrect. B does indicate rod journals.

Answer B is correct. C indicates main bearing journals, not the crankshaft thrust surface.

Answer C is incorrect. A does indicate oil supply holes.

Answer D is incorrect. C indicates main bearing journals.

50. Technician A says TTY bolts provide a more uniform clamping force when compared to conventional head bolts. Technician B says TTY bolts are tightened to a specified torque, then rotated a specified number of degrees. Who is correct?

TASK B.7

    A.   A only

    B.   B only

    C.   Both A and B

    D.   Neither A nor B

Answer A is incorrect. Technician B is also correct.

Answer B is incorrect. Technician A is also correct.

Answer C is correct. Both Technicians are correct. TTY bolts do provide more uniform clamping force, and they are rotated a specified angle after initial torque.

Answer D is incorrect. Both Technicians are correct.

PREPARATION EXAM 3—ANSWER KEY

1.	A	**21.**	D	**41.**	B
2.	C	**22.**	C	**42.**	C
3.	A	**23.**	D	**43.**	A
4.	A	**24.**	C	**44.**	B
5.	B	**25.**	C	**45.**	B
6.	B	**26.**	C	**46.**	C
7.	C	**27.**	D	**47.**	A
8.	D	**28.**	D	**48.**	C
9.	B	**29.**	B	**49.**	B
10.	C	**30.**	C	**50.**	B
11.	D	**31.**	C		
12.	D	**32.**	C		
13.	C	**33.**	C		
14.	B	**34.**	C		
15.	B	**35.**	C		
16.	B	**36.**	C		
17.	A	**37.**	C		
18.	A	**38.**	C		
19.	D	**39.**	D		
20.	C	**40.**	C		

PREPARATION EXAM 3—EXPLANATIONS

TASK D.5

1. Technician A says a reverse flow cooling system will warm an engine up quicker than conventional flow. Technician B says a reverse flow cooling system directs coolant to the block and then to the head. Who is correct?

   A. A only

   B. B only

   C. Both A and B

   D. Neither A nor B

   Answer A is correct. Only Technician A is correct. A reverse flow cooling system directs the coolant through the head, which is the hottest portion of the engine, first and then through the block; the engine warms up quicker and produces fewer emissions during warm-up.

   Answer B is incorrect. A reverse flow cooling system directs the coolant to the head and then to the block.

   Answer C is incorrect. Only Technician A is correct.

   Answer D is incorrect. Technician A is correct.

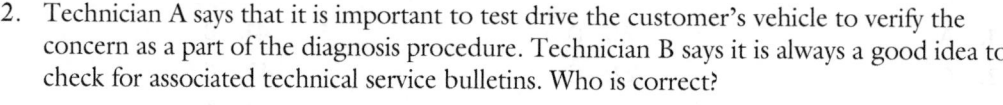

TASK A.1

2. Technician A says that it is important to test drive the customer's vehicle to verify the concern as a part of the diagnosis procedure. Technician B says it is always a good idea to check for associated technical service bulletins. Who is correct?

   A. A only

   B. B only

   C. Both A and B

   D. Neither A nor B

   Answer A is incorrect. Technician B is also correct.

   Answer B is incorrect. Technician A is also correct.

   Answer C is correct. Both Technicians are correct. A test drive lets you confirm the customer's complaint and checking for technical service bulletins may save diagnostic time.

   Answer D is incorrect. Both Technicians are correct.

TASK C.12

3. Technician A says that when measuring piston ring groove-to-ring clearance, you should place the ring into the groove and measure clearance with a feeler gauge. Technician B says the ring groove clearance is the same as the ring end gap clearance. Who is correct?

   A. A only

   B. B only

   C. Both A and B

   D. Neither A nor B

   Answer A is correct. Only Technician A is correct. The clearance between the ring and the ring land must be within manufacturer's specifications. Too wide a clearance will cause ring flutter and damage; too small a clearance and the ring will bind in the land and seize, causing a loss of compression.

   Answer B is incorrect. The clearance between the ring and ring land is usually between 0.0005 and 0.001 inch; ring end gap is much larger. Manufacturer's specifications should be checked for each engine to ensure proper fit.

   Answer C is incorrect. Only Technician A is correct.

   Answer D is incorrect. Technician A is correct.

4. Technician A says that valve springs must be pressure tested at both open height and closed height. Technician B says if the pressure is low at the closed height but good at the open height, the spring can be reused. Who is correct?

TASK B.4

   A. A only

   B. B only

   C. Both A and B

   D. Neither A nor B

Answer A is correct. Only Technician A is correct. The pressure must be tested at both open valve height and closed valve height to ensure the proper tension on the valve at all times.

Answer B is incorrect. The pressure must be within specifications at both test heights to properly control tension on the valve.

Answer C is incorrect. Only Technician A is correct.

Answer D is incorrect. Technician A is correct.

5. After performing a compression test on a 4-cylinder engine, the results show low pressure on cylinders #1 and #3. Technician A says this could be caused by a blown head gasket between the cylinders. Technician B says a cylinder leakage test would help pinpoint the problem. Who is correct?

TASK A.8

   A. A only

   B. B only

   C. Both A and B

   D. Neither A nor B

Answer A is incorrect. Cylinders #1 and #3 are not adjacent cylinders. The head gasket cannot be blown between the two.

Answer B is correct. A cylinder leakage test would help pinpoint the source of compression loss.

Answer C is incorrect. Only Technician B is correct.

Answer D is incorrect. Technician B is correct.

6. Technician A says a power balance test is used to pinpoint the cause of low compression test results. Technician B says a power balance test is used to pinpoint an underperforming cylinder. Who is correct?

TASK A.7

   A. A only

   B. B only

   C. Both A and B

   D. Neither A nor B

Answer A is incorrect. A cylinder leakage test would be used to pinpoint the cause of low compression.

Answer B is correct. Only Technician B is correct. During a power balance test, each cylinder is shorted one at a time. The cylinder with the least amount of RPM drop is the underperforming cylinder.

Answer C is incorrect. Only Technician B is correct.

Answer D is incorrect. Technician B is correct.

TASK E.5

7. Technician A says an oil leak at the rear main seal could be caused by worn rings. Technician B says an oil leak at the rear main seal could be caused by a defective PCV valve or hose. Who is correct?

   A.   A only

   B.   B only

   C.   Both A and B

   D.   Neither A nor B

   Answer A is incorrect. Technician B is also correct.

   Answer B is incorrect. Technician A is also correct.

   Answer C is correct. Both Technicians are correct. Worn rings can cause excessive blowby, creating pressure in the crankcase and allowing oil to leak from the rear main seal. A defective PCV system will do much the same thing.

   Answer D is incorrect. Both Technicians are correct.

TASK A.9

8. A technician is performing a cylinder leakage test and cylinders #3 and #4 have 65 percent leakage. Which is the LEAST LIKELY place the technician would look for escaping air?

   A.   The adjacent cylinder

   B.   The intake system

   C.   The exhaust system

   D.   The transmission dipstick

   Answer A is incorrect. If two adjacent cylinders have similar leakage, the head gasket may be blown between those cylinders.

   Answer B is incorrect. The intake system should be checked.

   Answer C is incorrect. The exhaust system should be checked.

   Answer D is correct. The transmission will not show air leakage from a cylinder in the engine.

TASK B.9

9. Technician A says the valve stem installed height is measured from the bottom of the valve guide to the top of the installed valve. Technician B says the valve stem installed height is measured from the spring seat to the tip of the installed valve. Who is correct?

   A.   A only

   B.   B only

   C.   Both A and B

   D.   Neither A nor B

   Answer A is incorrect. Valve stem installed height is measured from the valve seat to the tip of the valve.

   Answer B is correct. Only Technician B is correct. Valve stem installed height is measured from the valve seat to the tip of the valve.

   Answer C is incorrect. Only Technician B is correct

   Answer D is incorrect. Technician B is correct.

10. Technician A says a rod bearing which is worn will sound like a deep knock at the bottom of the engine. Technician B says a loose piston pin will cause a double click noise. Who is correct?

TASK A.4

    A. A only

    B. B only

    C. Both A and B

    D. Neither A nor B

Answer A is incorrect. Technician B is also correct.

Answer B is incorrect. Technician A is also correct.

Answer C is correct. Both Technicians are correct. Noise from a worn rod bearing will come from the bottom of the engine and be a deep knock. A loose piston pin will cause a double click as the piston reaches TDC or BDC and changes direction.

Answer D is incorrect. Both Technicians are correct.

11. If a block is to be rebuilt, all of the following parts should be removed during teardown EXCEPT:

TASK C.1

    A. Core (freeze) plugs.

    B. Oil galley plugs.

    C. In-block camshaft bearings.

    D. Cylinder liners.

Answer A is incorrect. The core plugs should be removed so the water jackets can be cleaned.

Answer B is incorrect. The oil galley plugs must be removed so the galleys can be cleaned of any old oil residue.

Answer C is incorrect. The camshaft bearings must be removed and replaced.

Answer D is correct. The cylinder liners are usually cast-in-place and are not removable except by boring.

12. The exhaust coming from the tailpipe is blue in color. Technician A says this is caused by the engine running too rich. Technician B says this is caused by coolant entering the combustion chamber. Who is correct?

TASK A.5

    A. A only

    B. B only

    C. Both A and B

    D. Neither A nor B

Answer A is incorrect. Black smoke from the tailpipe indicates a rich condition.

Answer B is incorrect. Coolant entering the combustion chamber will cause the exhaust to be white in color.

Answer C is incorrect. Neither Technician is correct.

Answer D is correct. Neither Technician is correct. Blue smoke from the tailpipe indicates oil being burned in the combustion chamber.

TASK E.7

13. The technician installs a vacuum gauge on the engine and notes a normal reading at idle, but when the RPM is raised, the needle fluctuates rapidly between 12 and 24 in. Hg. Which of the following is the most likely cause?

A. A leaking intake manifold

B. A misfiring spark plug

C. Weak valve springs

D. Weak rings

Answer A is incorrect. A leaking intake manifold will cause low vacuum at idle.

Answer B is incorrect. A misfiring spark plug would cause a rough idle and fluctuating vacuum.

Answer C is correct. Weak valve springs will cause slow or poor seating of the valves as the engine RPM increases, creating a fluctuating needle.

Answer D is incorrect. Weak rings will cause a low vacuum reading at idle.

TASK E.2

14. Technician A says on a vehicle with a map sensor, the air intake ductwork is not necessary. Technician B says that intact air intake ductwork is critical to engine longevity. Who is correct?

A. A only

B. B only

C. Both A and B

D. Neither A nor B

Answer A is incorrect. Even though a map-equipped engine does not measure airflow in the ductwork, the ductwork does affect how well the engine breathes at various speeds.

Answer B is correct. Only Technician B is correct. An air intake system that is intact will prevent dust, pollen, etc. from entering the engine and causing accelerated wear.

Answer C is incorrect. Only Technician B is correct.

Answer D is incorrect. Technician B is correct.

TASK D.2

15. An engine experiences very low oil pressure. Technician A says this may be caused by a stuck closed pressure relief valve. Technician B says low oil pressure will cause valve train rattle. Who is correct?

A. A only

B. B only

C. Both A and B

D. Neither A nor B

Answer A is incorrect. A stuck closed pressure relief valve will cause high oil pressure.

Answer B is correct. Only Technician B is correct. Low oil pressure will be noticed as a rattling, ticking noise from the valve train since this area is the last to receive pressurized oil.

Answer C is incorrect. Only Technician B is correct.

Answer D is incorrect. Technician B is correct.

16. Technician A says that hot tanking an engine block after boring is sufficient cleaning before reassembly. Technician B says that cleaning the bores and block with hot soapy water and a bore brush ensures that all machining particles are removed. Who is correct?

    A. A only

    B. B only

    C. Both A and B

    D. Neither A nor B

TASK C.16

Answer A is incorrect. The bore must be scrubbed with soap and water using a bore brush to remove machining particles.

Answer B is correct. Only Technician B is correct. The proper way to clean a block and its bores is with soap, water, and a bore brush. The block should then be treated with a protective coating to keep it from rusting.

Answer C is incorrect. Only Technician B is correct.

Answer D is incorrect. Technician B is correct.

17. Technician A says a running (dynamic) compression test is used to check cylinder breathing. Technician B says a running (dynamic) compression test is used to check cylinder sealing. Who is correct?

    A. A only

    B. B only

    C. Both A and B

    D. Neither A nor B

TASK A.8

Answer A is correct. Only Technician A is correct. A running compression test is used to check cylinder breathing and a cranking compression test is used to check cylinder sealing.

Answer B is incorrect. A cranking compression test is used to check cylinder sealing, not breathing.

Answer C is incorrect. Only Technician A is correct.

Answer D is incorrect. Technician A is correct.

18. Technician A says that incorrect valve adjustment may cause bent valves. Technician B says incorrect valve adjustment will not affect performance. Who is correct?

    A. A only

    B. B only

    C. Both A and B

    D. Neither A nor B

TASK B.12

Answer A is correct. Only Technician A is correct. Valves that are adjusted too tight may cause bent valves.

Answer B is incorrect. Incorrect valve adjustment will cause engine performance problems.

Answer C is incorrect. Only Technician A is correct.

Answer D is incorrect. Technician A is correct.

TASK C.5

19. Crankshaft inspection should include all of the following EXCEPT:

    A. Rod journal diameter and taper.
    B. Main journal diameter and taper.
    C. Dial indicator check for warpage.
    D. Crankshaft length.

    Answer A is incorrect. Rod journals should be checked for wear.

    Answer B is incorrect. Main journals should be checked for wear.

    Answer C is incorrect. The crankshaft should be checked for warpage.

    Answer D is correct. Overall crankshaft length will not change and does not need to be checked.

TASK E.6

20. Technician A says that a timing adjustment check is done with a timing light. Technician B says that timing adjustment may require a scan tool. Who is correct?

    A. A only
    B. B only
    C. Both A and B
    D. Neither A nor B

    Answer A is incorrect. Technician B is also correct.

    Answer B is incorrect. Technician A is also correct.

    Answer C is correct. Both Technicians are correct. Depending on the year of manufacture and equipment, the timing may be set using a timing light or through the ECM by a scan tool.

    Answer D is incorrect. Both Technicians are correct.

TASK A.8

21. The results of a cylinder cranking compression test on a 4-cylinder engine shows low compression on cylinder #3. The technician squirts a tablespoon of oil in the cylinder and retests. The compression reading increases by 40 percent. The increase in compression indicates:

    A. A leaking intake valve.
    B. A blown head gasket.
    C. A burned exhaust valve.
    D. Worn rings.

    Answer A is incorrect. A leaking intake valve will not seal better with oil in the cylinder.

    Answer B is incorrect. Oil in the cylinder will not seal a leaking head gasket.

    Answer C is incorrect. Oil put in the cylinder will not seal a burned exhaust valve.

    Answer D is correct. Oil put into the cylinder will temporarily seal the rings and cause an increase in compression readings.

22. A vehicle with 110,000 miles on it is brought to the shop to be checked for low oil pressure. Technician A says the oil pressure should first be checked with a mechanical gauge to eliminate the pressure sending unit as the problem. Technician B says low oil pressure can be caused by worn crankshaft main bearings. Who is correct?

TASK D.1

   A.   A only
   B.   B only
   C.   Both A and B
   D.   Neither A nor B

   Answer A is incorrect. Technician B is also correct.

   Answer B is incorrect. Technician A is also correct.

   Answer C is correct. Both Technicians are correct. Oil pressure should first be checked with a mechanical oil pressure gauge to eliminate the sending unit or the dash gauge as the problem. Worn main bearings will cause low, to extremely low, oil pressure.

   Answer D is incorrect. Both Technicians are correct.

23. An 8-cylinder block is being inspected and it is noted that cylinder #4 has a small crack in the middle of the bore. Technician A says that the cylinder can be sleeved to be repaired. Technician B says the crack can be welded and bored. Who is correct?

TASK C.2

   A.   A only
   B.   B only
   C.   Both A and B
   D.   Neither A nor B

   Answer A is incorrect. A cracked cylinder bore cannot be repaired with a sleeve; the block must be replaced.

   Answer B is incorrect. A cracked cylinder bore cannot be welded; the block must be replaced.

   Answer C is incorrect. Neither Technician is incorrect.

   Answer D is correct. Neither Technician is correct. A cracked cylinder bore cannot be repaired. The heat and pressure of combustion will cause the sleeve to shift in the bore and the welded bore to crack again. The block must be replaced.

24. The exhaust outlet of a turbocharger is coated with oil. Which of the following is the most likely cause?

TASK E.3

   A.   Leaking valve stem seals
   B.   A plugged PCV system
   C.   Leaking turbocharger oil seals
   D.   Worn engine rings

   Answer A is incorrect. Leaking valve stem seals would cause oil to be burned in the combustion chamber.

   Answer B is incorrect. A plugged PCV system can cause oil burning in the combustion chamber or external leaks.

   Answer C is correct. If the turbocharger oil seals are leaking, liquid oil can be drawn into the outlet of a turbocharger.

   Answer D is incorrect. Worn engine rings will cause poor performance and oil burning in the combustion chamber.

TASK B.13

25. Timing belt replacement is being discussed. What is the first thing that should be done?

  A. Remove the harmonic balancer.
  B. Remove the water pump.
  C. Align all timing marks.
  D. Release the timing belt tensioner.

Answer A is incorrect. The balancer would be removed after aligning the timing marks.

Answer B is incorrect. The water pump may not have to be removed.

Answer C is correct. All of the other choices may have to be done but the first step should be to align the timing marks.

Answer D is incorrect. This would only be done after aligning the timing marks.

TASK A.7

26. During a power balance test on a port fuel-injected engine, one cylinder is found to have virtually no RPM change. Which of these is the most likely cause?

  A. A faulty crankshaft position sensor
  B. A vacuum leak at the throttle body
  C. A defective plug wire
  D. A faulty camshaft position sensor

Answer A is incorrect. A faulty crankshaft position sensor will affect all cylinders rather than just one.

Answer B is incorrect. A vacuum leak at the throttle body will affect all cylinders.

Answer C is correct. If the plug wire on that cylinder is defective, the cylinder will not fire and contribute to power.

Answer D is incorrect. A faulty camshaft position sensor would affect all cylinders.

TASK C.4

27. Technician A says that a cylinder bore that has a deep scratch can be repaired using a dingleberry hone. Technician B says that a cylinder bore that has 0.008 inch taper does not need to be bored. Who is correct?

  A. A only
  B. B only
  C. Both A and B
  D. Neither A nor B

Answer A is incorrect. A deep scratch in the cylinder bore cannot be removed with a dingleberry hone; it will require a stone hone or boring.

Answer B is incorrect. A cylinder bore with 0.008 inch taper must be bored straight to ensure proper sealing by the rings.

Answer C is incorrect. Neither Technician is correct.

Answer D is correct. Neither Technician is correct. Both cylinder bore conditions will require the cylinder to be bored.

28. Technician A says that a 13 lb. radiator cap that is being replaced could be replaced with a 15 lb. cap for better protection. Technician B says a 13 lb. cap that releases pressure at 10 lbs. when being tested is OK to be used. Who is correct?

TASK D.6

    A.   A only

    B.   B only

    C.   Both A and B

    D.   Neither A nor B

Answer A is incorrect. A 13 lb. cap should only be replaced with a cap of the same rating.

Answer B is incorrect. A radiator cap should hold pressure until its rated pressure is exceeded.

Answer C is incorrect. Neither Technician is correct.

Answer D is correct. Neither Technician is correct. The radiator cap rating should not be changed; a 13 lb. cap replaced with a 15 lb. cap may cause cooling system leaks. A radiator cap which releases pressure before its rating may cause the engine to overheat. A radiator cap raises the boiling point of the coolant 3 degrees for every lb. of cap pressure. A cap that releases at 10 lb. rather than its rated 13 lb. will cause the coolant to boil 9 degrees before the proper pressure is reached.

29. Technician A says a stuck open thermostat can cause high pressure in the cooling system. Technician B says a defective radiator pressure cap can cause a burst upper radiator tank. Who is correct?

TASK D.9

    A.   A only

    B.   B only

    C.   Both A and B

    D.   Neither A nor B

Answer A is incorrect. A stuck open thermostat can cause low pressure rather than high.

Answer B is correct. Only Technician B is correct. A pressure cap that does not release at its specified pressure may cause pressure to build to the point that the radiator fails.

Answer C is incorrect. Only Technician B is correct.

Answer D is incorrect. Technician B is correct.

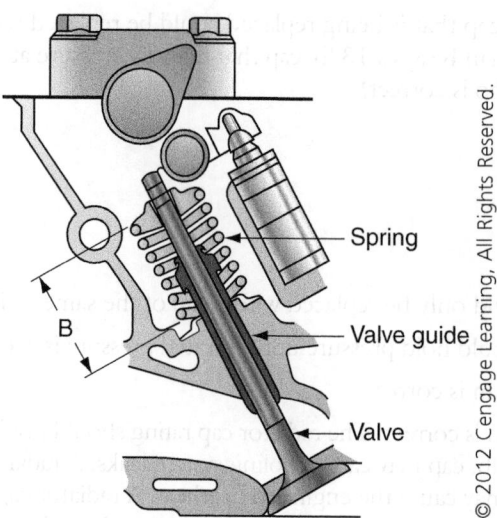

© 2012 Cengage Learning, All Rights Reserved.

Spring

Valve guide

Valve

B

TASK B.9

30. Measurement B in the figure above is more than specified. Technician A says it can be corrected by inserting a washer below the valve spring. Technician B says this could case improper valve seating. Who is correct?

   A. A only

   B. B only

   C. Both A and B

   D. Neither A nor B

   Answer A is incorrect. Technician B is also correct.

   Answer B is incorrect. Technician A is also correct.

   Answer C is correct. Both Technicians are correct. The spring installed height is corrected with spring washers. If the tension of the spring acting on the valve is not strong enough, the valve will close slower than the rate of cam lobe taper and the valve may not seat tightly against its seat.

   Answer D is incorrect. Both Technicians are correct.

TASK D.7

31. Technician A says a serpentine belt with three or more cracks within a one inch space should be replaced. Technician B says a smooth surface pulley that is uneven on the belt contact surface should be replaced. Who is correct?

   A. A only

   B. B only

   C. Both A and B

   D. Neither A nor B

   Answer A is incorrect. Technician B is also correct.

   Answer B is incorrect. Technician A is also correct.

   Answer C is correct. Both Technicians are correct. A serpentine belt with three or more cracks in a one inch span should be replaced. A smooth surface pulley that has worn unevenly will cause uneven pressure on the belt, causing it to slip or fail early.

   Answer D is incorrect. Both Technicians are correct.

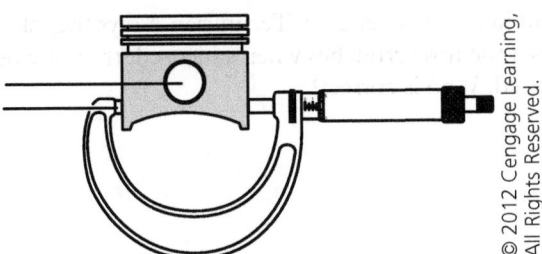

© 2012 Cengage Learning,
All Rights Reserved.

32. Technician A says that in the figure above, the piston diameter is being measured. Technician B says the piston should be measured on the skirt just above the bottom. Who is correct?

TASK C.10

    A.   A only

    B.   B only

    C.   Both A and B

    D.   Neither A nor B

Answer A is incorrect. Technician B is also correct.

Answer B is incorrect. Technician A is also correct.

Answer C is correct. Both Technicians are correct. The piston diameter is measured on the lower part of the skirt 90 degrees from the piston pin.

Answer D is incorrect. Both Technicians are correct.

33. An overhead cam engine has overheated and blown a head gasket. The head has been removed. All of the following would be inspected EXCEPT:

TASK B.2

    A.   Head deck for warpage.

    B.   Camshaft bore for warpage.

    C.   Camshaft followers for damage.

    D.   Head for cracks.

Answer A is incorrect. The deck surface would be checked for warpage.

Answer B is incorrect. If the deck was warped, the camshaft bore may also be warped and should be inspected.

Answer C is correct. The camshaft followers would not need to be inspected.

Answer D is incorrect. The head should be pressure tested for cracks which may have been caused by overheating, or may have been the cause for overheating.

34. Technician A says a pushrod engine's camshaft should be checked for lobe wear. Technician B says the camshaft should be checked for warping. Who is correct?

TASK B.13

    A.   A only

    B.   B only

    C.   Both A and B

    D.   Neither A nor B

Answer A is incorrect. Technician B is also correct.

Answer B is incorrect. Technician A is also correct.

Answer C is correct. Both Technicians are correct. A pushrod engine's camshaft should be checked for both warping and lobe wear.

Answer D is incorrect. Both Technicians are correct.

TASK D.8

35. The technician is checking the coolant hoses on an engine. Technician A says they should be checked for cracks. Technician B says if the hose crunches when squeezed, there are deposits inside the hose and it should be replaced. Who is correct?

A. A only

B. B only

C. Both A and B

D. Neither A nor B

Answer A is incorrect. Technician B is also correct.

Answer B is incorrect. Technician A is also correct.

Answer C is correct. Both Technicians are correct. Any deterioration of a coolant hose is grounds for replacement. Cracking or crunching when squeezed indicates the hose should be replaced immediately and the cooling system flushed.

Answer D is incorrect. Both Technicians are correct.

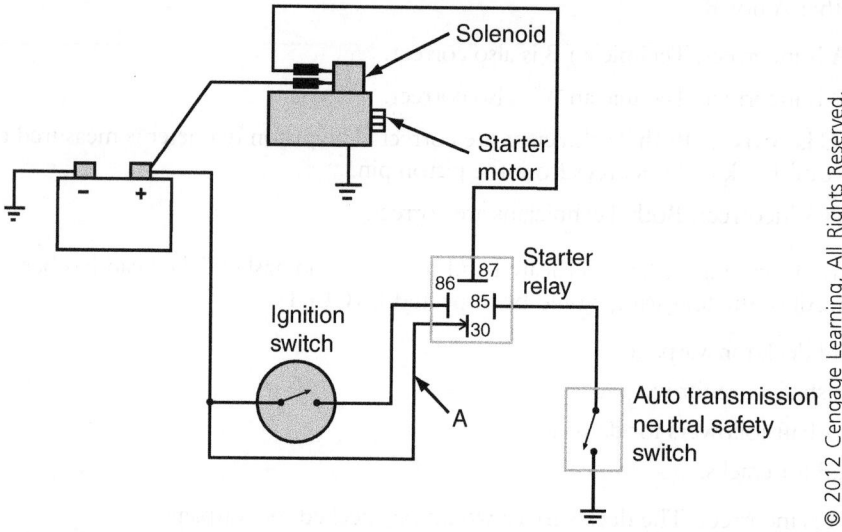

© 2012 Cengage Learning, All Rights Reserved.

TASK E.4

36. Technician A says an open in the wiring at point A in the figure above will prevent the starter from engaging. Technician B says if the neutral safety switch is open, the starter will not engage. Who is correct?

A. A only

B. B only

C. Both A and B

D. Neither A nor B

Answer A is incorrect. Technician B is also correct.

Answer B is incorrect. Technician A is also correct.

Answer C is correct. Both Technicians are correct. An open at point A will prevent current flow to the starter from the battery. An open at the neutral safety switch will cause the relay to stay open and will not close the current path from the battery to the starter.

Answer D is incorrect. Both Technicians are correct.

37. When a valve spring is checked for warpage, it is:

    A. Rolled on a flat surface.

    B. Compressed to see if it bows out.

    C. Placed on a flat surface standing up against a straightedge and rotated.

    D. Measured for free length.

TASK B.5

Answer A is incorrect. The spring is rotated next to a straightedge.

Answer B is incorrect. Compressing the spring may not show distortion.

Answer C is correct. Rotating the standing spring against a straightedge is the proper way to check a spring for warpage.

Answer D is incorrect. Spring free length is a separate spring measurement.

38. When the coolant is being tested, which of the following are the LEAST LIKELY to be tested for?

    A. Freeze protection

    B. pH level

    C. Flow volume

    D. Electrical current

TASK D.10

Answer A is incorrect. Freeze protection is always checked.

Answer B is incorrect. pH level should be checked. This test gives a better idea of the condition of the additives in the coolant.

Answer C is correct. Flow volume is not checked; that is a function of the water pump.

Answer D is incorrect. Checking for stray electrical current is recommended. Electrical current flow can cause rapid hose deterioration and reoccurring pinholes in the radiator and heater core.

39. Three types of valve stem seals are:

    A. Lip, o-ring, and floating.

    B. O-ring, positive, and box.

    C. Positive, umbrella, and lip.

    D. Positive, o-ring, and umbrella.

TASK B.6

Answer A is incorrect. Only answer D is correct.

Answer B is incorrect. Only answer D is correct.

Answer C is incorrect. Only answer D is correct.

Answer D is correct. The three types of valve stem seals in use are the positive lock, o-ring, and umbrella seal.

40. Which of the following would LEAST LIKELY require crankshaft grinding?

    A. An out of round journal

    B. Excessive journal taper

    C. Fine journal scoring

    D. Damaged thrust bearing surface

TASK C.5

Answer A is incorrect. An out of round journal must be trued.

Answer B is incorrect. Excessive journal taper would require grinding to correct.

Answer C is correct. Fine journal scoring can be polished with a special belt sander.

Answer D is incorrect. A damaged thrust bearing surface must be repaired.

TASK B.7

41. Technician A says a worn valve guide insert can be knurled to repair it. Technician B says a worn valve guide insert must be replaced. Who is correct?

    A. A only

    B. B only

    C. Both A and B

    D. Neither A nor B

Answer A is incorrect. A valve guide insert is a steel tube and is harder than the knurling tool; it will damage the tool.

Answer B is correct. Only Technician B is correct. A valve guide insert that is worn should be removed and replaced with a new insert.

Answer C is incorrect. Only Technician B is correct.

Answer D is incorrect. Technician B is correct.

TASK B.17

42. Cylinder heads are being reinstalled on the short block. Technician A says the head bolt holes should be chased with a tap before installation. Technician B says it is necessary to lubricate the head bolt threads before installation. Who is correct?

    A. A only

    B. B only

    C. Both A and B

    D. Neither A nor B

Answer A is incorrect. Technician B is also correct.

Answer B is incorrect. Technician A is also correct.

Answer C is correct. Both Technicians are correct. The head bolt holes should be chased with a tap to clean the threads and prevent incorrect torque. The head bolts should be lightly lubricated to prevent incorrect torque.

Answer D is incorrect. Both Technicians are correct.

TASK B.8

43. Technician A says that after a 45 degree valve seat has been cut, a 30 degree angle is used to position the valve-to-seat contact. Technician B says a 60 degree angle is used to position the valve-to-seat contact. Who is correct?

    A. A only

    B. B only

    C. Both A and B

    D. Neither A nor B

Answer A is correct. Only Technician A is correct. To adjust the valve-to-seat contact, a 30 degree angle is used for positioning.

Answer B is incorrect. A 60 degree angle is used to bottom cut the seat to narrow it if the seat is too wide.

Answer C is incorrect. Only Technician A is correct.

Answer D is incorrect. Technician A is correct.

44. An engine cranks over slowly and will not start. Technician A says the engine has to crank at 450 rpm to start. Technician B says the vehicle may have a weak battery. Who is correct?

TASK A.2

    A. A only
    B. B only
    C. Both A and B
    D. Neither A nor B

    Answer A is incorrect. An engine has to crank at 250 rpm minimum to start.

    Answer B is correct. Only Technician B is correct. If the vehicle's battery is weak, it will not supply the current needed to spin the engine at a minimum of 250 rpm.

    Answer C is incorrect. Only Technician B is correct.

    Answer D is incorrect. Technician B is correct.

45. There is coolant lost from an engine but the technician cannot tell where the leak is. Technician A says to pressurize the cooling system to 25 psi, then check for leaks. Technician B says small leaks may only be obvious immediately after a hot shutdown. Who is correct?

TASK D.3

    A. A only
    B. B only
    C. Both A and B
    D. Neither A nor B

    Answer A is incorrect. A cooling system should not be pressurized over the radiator cap rating.

    Answer B is correct. Only Technician B is correct. When an engine is shut down after reaching operating temperature, the pressure in the cooling system increases. This may cause the leak to become obvious.

    Answer C is incorrect. Only Technician B is correct.

    Answer D is incorrect. Technician B is correct.

46. During a vacuum test, it is noted that the needle oscillates rapidly as the engine RPM increases but it is steady and normal at idle. What would be the most likely cause?

TASK A.6

    A. Worn rings
    B. Sticky valves
    C. Weak valve springs
    D. A clogged exhaust system

    Answer A is incorrect. Worn rings will cause a low vacuum reading since the cylinder is not sealing properly.

    Answer B is incorrect. Sticky valves will show up as a steady and normal vacuum reading with occasional sharp flicks down.

    Answer C is correct. Weak valve springs will cause rapid needle oscillation as RPM increases. The valves will be slow to close causing poor cylinder evacuation.

    Answer D is incorrect. A clogged exhaust system will cause low vacuum readings when the RPM is held steady above 1,800 rpm, but at idle, the readings may be normal.

© 2012 Cengage Learning, All Rights Reserved.

TASK C.7

47. Technician A says the figure above shows the two halves of a one-piece main and thrust bearing. Technician B says there are two of these in an engine. Who is correct?

    A. A only

    B. B only

    C. Both A and B

    D. Neither A nor B

Answer A is correct. Only Technician A is correct. The figure shows a one-piece main and thrust bearing.

Answer B is incorrect. There will be only one thrust bearing set per engine.

Answer C is incorrect. Only Technician A is correct.

Answer D is incorrect. Technician A is correct.

TASK B.17

48. The technician is preparing to install the cylinder head on a 4-cylinder OHC engine with the cam installed. Which of these operations is the LEAST LIKELY to be performed?

    A. Run a bottoming tap through the head bolt holes.

    B. Apply a light coating of oil on the head bolts and their washers.

    C. Put #1 piston at TDC.

    D. Clean the deck surfaces of any oil residue.

Answer A is incorrect. Head bolt holes should be chased with a tap to ensure clean threads.

Answer B is incorrect. A light coating of oil allows the proper torque to be reached without binding and creating undertorque.

Answer C is correct. If the piston is at TDC when installing the head, there is a chance the valves will be damaged.

Answer D is incorrect. All deck surfaces must be clean to prevent contamination of the head gasket.

49. The threads in a water pump mounting hole have been damaged. Technician A says the threads may be drilled out oversize and tapped with the proper tap size and thread pitch to accept a larger diameter bolt. Technician B says the threads may be restored using a HeliCoil thread insert. Who is correct?

    TASK B.3

    A. A only

    B. B only

    C. Both A and B

    D. Neither A nor B

    Answer A is incorrect. While a hole can be drilled oversize for a larger diameter bolt, the water pump may not have enough surface area around the corresponding hole to be drilled oversize. This would not be the preferred method to repair the threads.

    Answer B is correct. Only Technician B is correct. The preferred method to repair the threads would be to use a HeliCoil thread insert to renew the threads to the original size and thread pitch.

    Answer C is incorrect. Only Technician B is correct.

    Answer D is incorrect. Technician B is correct.

50. A running (dynamic) compression test has been performed after a cranking compression test was done on a 4-cylinder engine. The results indicated cylinder #3 has substantially lower running compression than the other three. Cranking compression was within +/− 10 percent for all four cylinders. The running compression test results for cylinder #3 indicates:

    TASK A.8

    A. A restriction in the exhaust for that cylinder.

    B. A restriction in the intake air on that cylinder.

    C. A stopped-up catalytic converter.

    D. A plugged air filter.

    Answer A is incorrect. A restriction in the exhaust flow from that cylinder would create a higher running compression test results. A sticky valve or restriction in the exhaust port for cylinder #3 would not allow the emptying of the cylinder, and the next intake stroke would increase the pressure in that cylinder.

    Answer B is correct. A restriction of intake for cylinder #3 would cause the running compression test results to be much lower as that cylinder is not ingesting as much air on the intake stroke.

    Answer C is incorrect. A stopped-up catalytic converter would cause higher than expected running test compression results since none of the cylinders would empty properly.

    Answer D is incorrect. A plugged air filter would restrict air intake to all cylinders, not just #3.

PREPARATION EXAM 4—ANSWER KEY

1.	A	21.	C	41.	C
2.	C	22.	A	42.	B
3.	D	23.	C	43.	D
4.	D	24.	D	44.	B
5.	B	25.	A	45.	C
6.	B	26.	C	46.	A
7.	B	27.	A	47.	B
8.	A	28.	C	48.	D
9.	B	29.	D	49.	C
10.	C	30.	D	50.	D
11.	D	31.	D		
12.	B	32.	B		
13.	D	33.	B		
14.	A	34.	B		
15.	B	35.	D		
16.	A	36.	B		
17.	D	37.	C		
18.	C	38.	A		
19.	C	39.	D		
20.	C	40.	C		

PREPARATION EXAM 4—EXPLANATIONS

TASK D.1

1. The customer says his oil pressure gauge reads low even at highway speeds. Technician A says this may be caused by worn crankshaft main bearings. Technician B says this can be caused by leaking rings. Who is correct?

   A. A only

   B. B only

   C. Both A and B

   D. Neither A nor B

   Answer A is correct. Only Technician A is correct. Worn main bearings will cause oil pressure to be low at all speeds.

   Answer B is incorrect. Leaking rings will cause compression loss but will not affect oil pressure.

   Answer C is incorrect. Only Technician A is correct.

   Answer D is incorrect. Technician A is correct.

2. The technician is diagnosing an intermittent no-crank problem. Which of the following would be the LEAST LIKELY to cause this problem?

    A. A poor connection at the battery positive post

    B. High resistance in the starter ground circuit

    C. Hydro-locked engine

    D. A worn ignition switch

TASK A.2

Answer A is incorrect. A poor connection at the battery positive post can cause an intermittent no-crank condition due to high resistance.

Answer B is incorrect. High resistance in the starter ground circuit can cause intermittent no-crank conditions.

Answer C is correct. If the engine was hydro-locked, it would not be an intermittent problem; it would be a consistent no-crank.

Answer D is incorrect. A worn ignition switch can cause intermittent continuity to the starting circuit.

3. When preparing an engine for removal, which of the following would be done?

    A. Drain engine coolant.

    B. Drain engine oil.

    C. Disconnect fuel lines.

    D. All of the above

TASK C.1

Answer A is incorrect. Draining engine coolant, oil, and disconnecting the fuel lines would all be done.

Answer B is incorrect. Draining engine coolant, oil, and disconnecting the fuel lines would all be done.

Answer C is incorrect. Draining engine coolant, oil, and disconnecting the fuel lines would all be done.

Answer D is correct. All the fluids should be removed from the engine and the fuel lines disconnected.

4. The customer with a 1994 model vehicle complains of a strong fuel odor when he walks past the front of the car. Which of the following would be the LEAST LIKELY cause?

    A. Leaking injector seals

    B. A saturated evaporative emissions charcoal canister

    C. Cracked fuel line o-rings

    D. A leaking fuel pump

TASK A.3

Answer A is incorrect. Leaking injector seals could cause a fuel leak on the intake manifold.

Answer B is incorrect. An evaporative emissions charcoal canister may become saturated with raw fuel if the fuel tank is consistently overfilled. Many early canisters were located under the hood.

Answer C is incorrect. Leaking fuel line o-rings at the engine will cause a fuel leak under the hood.

Answer D is correct. By 1994, all engines were fuel-injected and the electric fuel pump was in the gas tank; a fuel pump leak could not cause a fuel odor.

TASK D.10

5. Technician A says extended life coolants provide rust and freeze protection for 5 years/ 150,000 miles. Technician B says extended life coolants provide rust and corrosion protection for 5 years/150,000 miles. Who is correct?

A. A only

B. B only

C. Both A and B

D. Neither A nor B

Answer A is incorrect. Extended life coolants provide only rust and corrosion protection for 5 years/150,000 miles, not freeze protection.

Answer B is correct. Only Technician B is correct. Extended life coolants only guarantee rust and corrosion protection, not freeze protection.

Answer C is incorrect. Only Technician B is correct.

Answer D is incorrect. Technician B is correct.

TASK A.8

6. The technician is performing a cylinder cranking compression test. Technician A says a variance of 35 percent is acceptable. Technician B says there should be no more than 20 percent between the highest and lowest cylinder results. Who is correct?

A. A only

B. B only

C. Both A and B

D. Neither A nor B

Answer A is incorrect. Most manufacturers allow a 20 percent difference in compression test results.

Answer B is correct. Only Technician B is correct. Most manufacturers allow a 20 percent difference in compression test results.

Answer C is incorrect. Only Technician B is correct.

Answer D is incorrect. Technician B is correct.

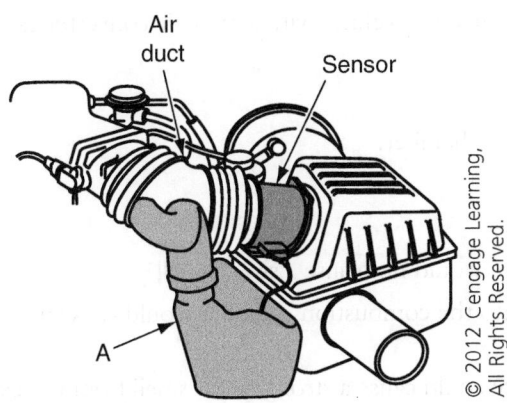

Air duct

Sensor

A

© 2012 Cengage Learning,
All Rights Reserved.

7. In the figure above, Technician A says A is a water separator. Technician B says it is a resonator to reduce intake moan. Who is correct?

    A. A only
    B. B only
    C. Both A and B
    D. Neither A nor B

TASK E.2

Answer A is incorrect. There should not be enough moisture in the intake to require a separator.

Answer B is correct. Only Technician B is correct. A resonator in the intake air duct reduces intake moan caused by air being sucked into the engine.

Answer C is incorrect. Only Technician B is correct.

Answer D is incorrect. Technician B is correct.

8. After doing a compression test, it is determined that an engine has a blown head gasket. Which of the following is the LEAST LIKELY result?

    A. Higher than normal compression readings
    B. Oil that is a milky brown color
    C. Bubbles in the radiator
    D. White exhaust smoke

TASK A.8

Answer A is correct. A blown head gasket will cause compression to be low, not high.

Answer B is incorrect. Coolant mixing with the oil will cause the oil to look milky.

Answer C is incorrect. A head gasket leak that allows combustion gases to enter the cooling system will cause bubbles in the radiator.

Answer D is incorrect. A head gasket leak that allows coolant to enter the combustion chamber will cause the exhaust to be white.

TASK A.5

9. A sweet smell and white smoke in the exhaust of a vehicle with a catalytic converter can be an indication of:

   A.   A lean fuel mixture.

   B.   Coolant leaking into the combustion chamber.

   C.   A rich fuel mixture.

   D.   A vacuum leak.

   Answer A is incorrect. A lean fuel mixture would not cause a sweet smell.

   Answer B is correct. Coolant leaking into the combustion chamber would cause white exhaust color with a sweet smell.

   Answer C is incorrect. A rich fuel mixture could cause a strong sulfur smell from the exhaust as the extra fuel is burned in the converter.

   Answer D is incorrect. A vacuum leak would cause a rough idle that would decrease as engine speed increases.

TASK D.11

10. Technician A says coolant with rust in it may cause the water pump impeller to fail. Technician B says the water pump impeller may be made of plastic. Who is correct?

    A.   A only

    B.   B only

    C.   Both A and B

    D.   Neither A nor B

    Answer A is incorrect. Technician B is also correct.

    Answer B is incorrect. Technician A is also correct.

    Answer C is correct. Both Technicians are correct. Rust in the coolant is an abrasive which may wear the impeller vanes off, causing little or no coolant circulation. Some water pumps in use today are made with plastic impellers.

    Answer D is incorrect. Both Technicians are correct.

TASK A.5

11. The catalytic converter is overheating causing it to glow bright red and heat the passenger's floorboard. There is also a strong sulfur odor from the exhaust. The cause of this could be:

    A.   Improper coolant in the engine.

    B.   A vacuum leak.

    C.   Too low a fuel octane.

    D.   A rich fuel mixture.

    Answer A is incorrect. Improper coolant will not affect the converter.

    Answer B is incorrect. A vacuum leak will not cause an exhaust odor.

    Answer C is incorrect. The wrong fuel octane may cause spark knock but will not cause the converter to overheat.

    Answer D is correct. Too rich a fuel mixture can cause the excess fuel to burn in the converter and overheat it, causing a strong sulfur odor.

12. Technician A says a lower-than-normal idle speed in a fuel-injected engine may be caused by a broken vacuum line. Technician B says an intake manifold vacuum leak may cause a misfire at idle and lower engine speeds. Who is correct?

TASK E.1

   A. A only
   B. B only
   C. Both A and B
   D. Neither A nor B

   Answer A is incorrect. A broken vacuum line may cause a higher-than-normal idle speed as the ECM adds fuel to correct a lean condition in the engine.

   Answer B is correct. Only Technician B is correct. A broken vacuum line may cause misfire at idle or low RPM. The vacuum leak introduces unmetered air into the intake, causing a lean condition in the cylinder and a misfire.

   Answer C is incorrect. Only Technician B is correct.

   Answer D is incorrect. Technician B is correct.

13. During a vacuum test, it is noted that when the engine is accelerated, the vacuum drops to near zero, then climbs back to the normal level. Technician A says that the piston rings are worn. Technician B says this indicates weak valve springs. Who is correct?

TASK A.6

   A. A only
   B. B only
   C. Both A and B
   D. Neither A nor B

   Answer A is incorrect. Worn piston rings will cause a low but steady vacuum, usually between 8 and 10 in. Hg.

   Answer B is incorrect. Weak valve springs will cause erratic needle fluctuations when the engine is accelerated smoothly.

   Answer C is incorrect. Neither Technician is correct.

   Answer D is correct. Neither Technician is correct. A vacuum drop to near zero and a return to normal indicates proper operation.

14. Technician A says that some RTV sealants may be harmful to an oxygen sensor. Technician B says that RTV is an anaerobic sealer. Who is correct?

TASK C.16

   A. A only
   B. B only
   C. Both A and B
   D. Neither A nor B

   Answer A is correct. Only Technician A is correct. Some older formula RTV fumes could damage the oxygen sensor. New formulas will not damage the oxygen sensor, and should say so on the packaging.

   Answer B is incorrect. RTV is an aerobic sealer; it requires air to cure.

   Answer C is incorrect. Only Technician A is correct.

   Answer D is incorrect. Technician A is correct.

TASK A.4

15. A front-wheel drive vehicle's engine jumps during acceleration and hits the underside of the hood. The motor mounts are inspected and although none of them appear to be broken, the front mount is coated with oil. Technician A says the motor mount was intact and was not the cause of engine movement. Technician B says the oil came from the fluid-filled hydraulic motor mount, which would allow excessive engine movement. Who is correct?

A. A only

B. B only

C. Both A and B

D. Neither A nor B

Answer A is incorrect. The front mount is usually broken if the engine rocks enough during acceleration, causing the engine to hit the hood. The front mount is often a fluid-filled hydraulic mount and, if the oil has leaked out, makes it ineffective.

Answer B is correct. Only Technician B is correct. Fluid-filled hydraulic engine mounts are becoming commonplace. If the fluid leaks out of the mount, it will not control engine movement and on some vehicles, the engine can hit the underside of the hood.

Answer C is incorrect. Only Technician B is correct.

Answer D is incorrect. Technician B is correct.

TASK E.3

16. Reduced turbocharger boost pressure may be caused by a:

A. Wastegate valve stuck open.

B. Leaking wastegate diaphragm.

C. Disconnected wastegate linkage.

D. Wastegate stuck closed.

Answer A is correct. A stuck open wastegate will bypass all exhaust from the turbo wheel.

Answer B is incorrect. A leaking wastegate diaphragm will apply excessive boost.

Answer C is incorrect. Disconnected wastegate linkage will allow constant boost.

Answer D is incorrect. A stuck closed wastegate will apply excessive boost.

TASK D.8

17. A heater hose is being replaced and the technician finds it is stuck on the hose nipple. Which of the following is the best solution to the problem?

A. Use pliers to twist the hose off.

B. Run a screwdriver between the hose and the nipple and pry it off.

C. Twist the hose by hand and pull on it at the same time.

D. Slice the hose lengthwise and peel it off.

Answer A is incorrect. Using pliers to twist the hose off will damage the hose nipple, making it difficult to seal a new hose.

Answer B is incorrect. The screwdriver will also damage the hose nipple.

Answer C is incorrect. Twisting the hose by hand may break the hose nipple from the heater core. Heater cores are easily damaged.

Answer D is correct. A stuck hose should be split and peeled off to avoid damaging the hose nipple or the heater core.

18. The technician is testing for a restricted catalytic converter. Technician A says a vacuum gauge may be used. Technician B says a temperature sensing probe can be used. Who is correct?

TASK E.7

   A. A only
   B. B only
   C. Both A and B
   D. Neither A nor B

Answer A is incorrect. Technician B is also correct.

Answer B is incorrect. Technician A is also correct.

Answer C is correct. Both Technicians are correct. A vacuum gauge will show very low vacuum on an engine with a restricted catalytic converter when the engine is run at 2,000 rpm. A temperature sensing probe, which is used to measure inlet and outlet temperature of the catalytic converter, can indicate a restricted converter. The normal outlet temperature should be about 10 percent higher than the inlet.

Answer D is incorrect. Both Technicians are correct.

19. After a vehicle sits overnight, it has a light tapping noise when it is first started that disappears after a short time. The most likely cause would be:

TASK A.4

   A. Low oil level.
   B. Worn rod bearings.
   C. Excessive lifter leakdown.
   D. Weak oil pump.

Answer A is incorrect. If the oil level was low, the noise would not go away.

Answer B is incorrect. If the rod bearings were making noise, it would be a deeper knock and probably would not go away.

Answer C is correct. If the lifters leakdown after sitting, when the oil pressure comes up they will pump up quickly and quiet down.

Answer D is incorrect. If the noise was caused by a weak oil pump, it would not go away.

20. Technician A says a running (dynamic) compression test should be done with the spark disabled and the fuel injector for that cylinder unplugged. Technician B says that a running compression tests results at idle should be about 50 percent of cranking compression results. Who is correct?

TASK A.8

   A. A only
   B. B only
   C. Both A and B
   D. Neither A nor B

Answer A is incorrect. Technician B is also correct.

Answer B is incorrect. Technician A is also correct.

Answer C is correct. Both Technicians are correct. A running compression test is done with spark and fuel disabled for the cylinder being tested. The compression tester will go into the spark plug hole so the coil or plug wire must be grounded or unplugged. The injector must be disabled to prevent fuel wash in the cylinder. Idle results for a running compression test should be about 50 percent of cranking compression results.

Answer D is incorrect. Both Technicians are correct.

TASK C.11

21. When the rod bearings on an engine were inspected, it was noted that one rod bearing set had wear on one side of the lower bearing and on the opposite side of the upper bearing. This could be caused by:

   A. Lack of lubrication.
   B. A tapered rod journal.
   C. A twisted connecting rod.
   D. A warped crankshaft.

   Answer A is incorrect. Lack of lubrication would cause wear on all the bearings.

   Answer B is incorrect. A tapered rod journal would affect both halves of the bearing in the same places.

   Answer C is correct. A twisted connecting rod will cause pressure on one side of the bottom bearing and the other side of the top bearing.

   Answer D is incorrect. A warped crankshaft will affect the main bearing wear.

TASK A.7

22. A cylinder power balance test is performed on a rough running 1998 model 8-cylinder engine with four coil packs. Two cylinders on opposite heads show no RPM drop. Technician A says the ignition coil for those two cylinders may be defective. Technician B says a camshaft out of sync with the crankshaft can cause this. Who is correct?

   A. A only
   B. B only
   C. Both A and B
   D. Neither A nor B

   Answer A is correct. Only Technician A is correct. If the non-performing cylinders are operated from the same coil pack, a defective coil pack could be the problem.

   Answer B is incorrect. A camshaft out of sync with the crankshaft can cause a rough running engine but the problem will show up on multiple cylinders.

   Answer C is incorrect. Only Technician A is correct.

   Answer D is incorrect. Technician A is correct.

TASK E.2

23. An air filter that is plugged with debris could cause all of the following EXCEPT:
   A. Poor performance.
   B. Sluggish acceleration.
   C. Higher fuel economy.
   D. Lower fuel economy.

   Answer A is incorrect. A plugged air filter can cause poor performance.

   Answer B is incorrect. A plugged air filter will cause sluggish acceleration.

   Answer C is correct. A plugged air filter will cause lower than normal fuel economy, not higher.

   Answer D is incorrect. A plugged air filter will cause lower fuel economy.

24. Valve springs are being inspected. All of the following would be checked EXCEPT:

    A. Spring free height is checked.
    B. Checked for open tension.
    C. Checked for closed tension.
    D. Spring diameter is checked.

    TASK B.9

    Answer A is incorrect. Spring free height is checked.

    Answer B is incorrect. Open tension is checked.

    Answer C is incorrect. Closed tension is checked.

    Answer D is correct. Spring diameter is not a value that would be checked.

25. The ring gear teeth of a flywheel are badly worn in multiple spots. Technician A says this could cause poor starter drive gear engagement. Technician B says this could cause a lower than normal starter current draw. Who is correct?

    A. A only
    B. B only
    C. Both A and B
    D. Neither A nor B

    TASK E.4

    Answer A is correct. Only Technician A is correct. Missing or worn teeth on a flywheel ring gear may cause drive gear skip or bind.

    Answer B is incorrect. If the starter is engaged or binding, the current draw would be normal or higher than normal.

    Answer C is incorrect. Only Technician A is correct.

    Answer D is incorrect. Technician A is correct.

26. A cylinder leakage test has been performed and the results were:

Cyl. #1	Cyl. #2	Cyl. #3	Cyl. #4
15%	95%	95%	10%

    Technician A says these results indicate a blown head gasket. Technician B says Cyl. #1 and Cyl. #4 are within acceptable limits. Who is correct?

    A. A only
    B. B only
    C. Both A and B
    D. Neither A nor B

    TASK A.9

    Answer A is incorrect. Technician B is also correct.

    Answer B is incorrect. Technician A is also correct.

    Answer C is correct. Both Technicians are correct. Two adjacent cylinders that show excessive leakage indicate there is probably a blown head gasket between cylinders #2 and #3. The acceptable limit for cylinder leakage is 20 percent, so cylinders #1 and #4 are within leakage limits.

    Answer D is incorrect. Both Technicians are correct.

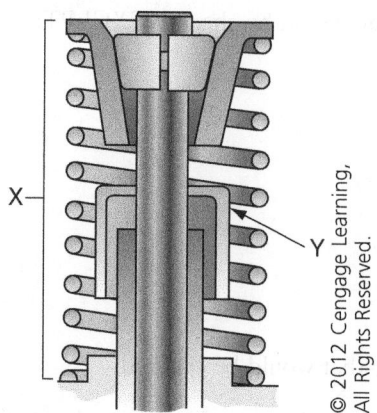

© 2012 Cengage Learning,
All Rights Reserved.

TASK B.6

27. Technician A says that X in the figure above can be replaced without removing the head. Technician B says the head must be removed to replace Y in the figure. Who is correct?

A. A only

B. B only

C. Both A and B

D. Neither A nor B

Answer A is correct. Only Technician A is correct. The valve spring can be removed, after filling the cylinder with air to hold the valve in place, and removing the keepers.

Answer B is incorrect. Y is the valve stem seal and can be removed and replaced using the same procedure as used for valve spring removal.

Answer C is incorrect. Only Technician A is correct.

Answer D is incorrect. Technician A is correct.

TASK A.5

28. A vehicle is being checked for coolant loss. There is no sign of coolant loss in the engine compartment. Technician A says coolant loss could be caused by a bad radiator cap. Technician B says the heater core may be leaking. Who is correct?

A. A only

B. B only

C. Both A and B

D. Neither A nor B

Answer A is incorrect. Technician B is also correct.

Answer B is incorrect. Technician A is also correct.

Answer C is correct. Both Technicians are correct. A bad radiator cap will not allow pressure to build in the system and may cause boil over. A heater core that is leaking, may leak onto the floorboards of the passenger compartment and not show up under the hood.

Answer D is incorrect. Both Technicians are correct.

29. The vehicle cranks normally but is consistently slow to start. Which of the following is the LEAST LIKELY cause?

TASK E.1

   A. A partially plugged fuel filter
   B. A failed fuel pump check valve
   C. Leaking injectors
   D. A defective fuel pressure regulator

Answer A is incorrect. A partially plugged fuel filter would restrict fuel flow to the injectors.

Answer B is incorrect. A defective fuel pump check valve will allow the fuel in the lines to return to the tank after shutdown. The pump will have to fill the lines before fuel pressure can be established.

Answer C is incorrect. Leaking injectors could cause flooding or wash cylinder walls with fuel, causing low compression and extended start time.

Answer D is correct. A defective fuel pressure regulator will not allow fuel to return to the tank. This could cause performance problems but not consistently extended start time.

30. Technician A says that excessive wear on the lower halves of the main bearing inserts is caused by prolonged idle operation. Technician B says this wear can be caused by prolonged high RPM. Who is correct?

TASK C.5

   A. A only
   B. B only
   C. Both A and B
   D. Neither A nor B

Answer A is incorrect. Prolonged idle operation will not cause this kind of wear as there is little stress on the crankshaft during idle operation.

Answer B is incorrect. Prolonged high RPM will not cause this kind of wear as oil pressure is high.

Answer C is incorrect. Both Technicians are incorrect.

Answer D is correct. Both Technicians are incorrect. Engine overloading, the higher loading at too low an RPM, will cause the crankshaft to squeeze out the oil film from the lower bearing half.

31. A cylinder cranking compression test has been done; the results were all well below manufacturer's specifications. Technician A says a wet cylinder compression test should be done to eliminate valves as the possible cause. Technician B says a power balance test should be done to identify the problem. Who is correct?

TASK A.8

   A. A only
   B. B only
   C. Both A and B
   D. Neither A nor B

Answer A is incorrect. A wet compression test is not used to check for valves that are not sealing. This test checks for ring leakage. If the results were substantially more after the wet compression, test it would indicate worn rings.

Answer B is incorrect. A power balance test is used to identify non-performing cylinders. The next test should be the wet cylinder compression test.

Answer C is incorrect. Neither Technician is correct.

Answer D is correct. Neither Technician is correct. A wet cylinder compression test is used to check for poorly sealing rings, and the power balance test is used to identify the cylinder(s) that have poor power production in comparison with the other cylinders.

TASK B.17

32. The head has been sent to the machine shop to be resurfaced. Technician A says that a belt sander will true the surface and smooth it. Technician B says the surface finish is critical to proper head gasket sealing and life. Who is correct?

   A. A only

   B. B only

   C. Both A and B

   D. Neither A nor B

Answer A is incorrect. A belt sander would give the surface a course finish and might cause the head thickness to vary from one end to the other.

Answer B is correct. Only Technician B is correct. The surface finish is critical to proper sealing and life of a head gasket. Different expansion rates between a cast iron block and an aluminum head will cause a head gasket to fail if the surfaces are not the proper finish.

Answer C is incorrect. Only Technician B is correct.

Answer D is incorrect. Technician B is correct.

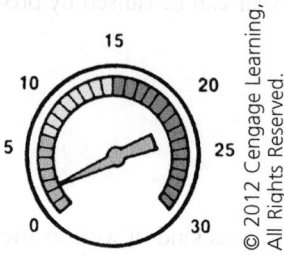

© 2012 Cengage Learning, All Rights Reserved.

TASK A.6

33. A vacuum test has been done and the results at 2,500 rpm are shown in the figure above. Technician A says the very low readings could be caused by a sticky valve. Technician B says the low, steady readings are probably the result of a severely restricted exhaust system. Who is correct?

   A. A only

   B. B only

   C. Both A and B

   D. Neither A nor B

Answer A is incorrect. A sticky valve would cause a normal reading that intermittently flicks down and comes right back.

Answer B is correct. Only Technician B is correct. A severely restricted exhaust system can cause these readings. The exhaust gases are not able to exit the combustion chamber, resulting in low vacuum readings.

Answer C is incorrect. Only Technician B is correct.

Answer D is incorrect. Technician B is correct.

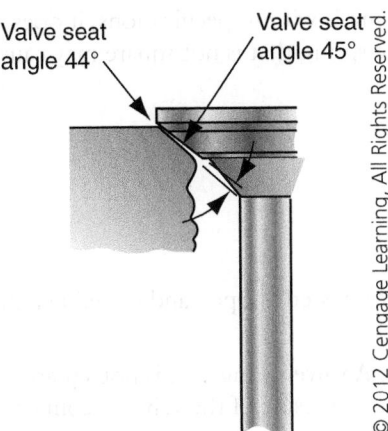

Valve seat angle 44°   Valve seat angle 45°

© 2012 Cengage Learning, All Rights Reserved.

34. In the figure above, Technician A says seat width is being shown. Technician B says the interference angle between the seat and the valve face is being shown. Who is correct?

TASK B.8

   A.   A only
   B.   B only
   C.   Both A and B
   D.   Neither A nor B

   Answer A is incorrect. Valve-to-seat interference angle is being shown.

   Answer B is correct. Only Technician B is correct. Valve-to-seat interference angle is being shown. Interference angle, if specified, is usually a one degree difference.

   Answer C is incorrect. Only Technician B is correct.

   Answer D is incorrect. Technician B is correct.

35. Which of the following steps is a technician LEAST LIKELY to perform when pressing the wrist pin into the piston and connecting rod?

TASK C.12

   A.   Align the bores in the piston and connecting rod.
   B.   Heat the small end of the rod.
   C.   Make sure the position marks on the piston and connecting rod are oriented properly.
   D.   Heat the wrist pin.

   Answer A is incorrect. The piston and connecting rod must be lined up to install the piston pin.

   Answer B is incorrect. The small end of the rod is heated to expand it for easy installation.

   Answer C is incorrect. The alignment marks for the piston and connecting rod must be aligned to ensure piston orientation in the cylinder bore.

   Answer D is correct. If the wrist pin is heated, it expands and will not go into the connecting rod hole.

TASK B.5

36. Technician A says if the free length of the valve spring is within specifications, it does not need to be tension tested. Technician B says a valve spring that is not square may cause uneven valve seat wear. Who is correct?

A. A only

B. B only

C. Both A and B

D. Neither A nor B

Answer A is incorrect. A valve spring must be tension tested at open and closed height; free length is only one of the measurements.

Answer B is correct. Only Technician B is correct. A valve spring that is not square can cause uneven seat wear. The warped spring will cause one side of the valve to contact with more pressure than the other side.

Answer C is incorrect. Only Technician B is correct.

Answer D is incorrect. Technician B is correct.

TASK B.14

37. Technician A says the exhaust lobes on a flat tappet lifter camshaft may wear more severely than the intake lobes. Technician B says this is due to increased pressure during valve opening. Who is correct?

A. A only

B. B only

C. Both A and B

D. Neither A nor B

Answer A is incorrect. Technician B is also correct.

Answer B is incorrect. Technician A is also correct.

Answer C is correct. Both Technicians are correct. The exhaust valve opens during the end of the combustion stroke when there is some pressure present in the cylinder, and can cause severe exhaust lobe wear on a flat tappet lifter engine.

Answer D is incorrect. Both Technicians are correct.

TASK C.5

38. Technician A says the crankshaft keyway should be inspected for chipping and wear. Technician B says a crankshaft that is cracked may be welded and reused. Who is correct?

A. A only

B. B only

C. Both A and B

D. Neither A nor B

Answer A is correct. Only Technician A is correct. The keyway should be inspected for damage, which will allow the balancer to shift.

Answer B is incorrect. A cracked crankshaft must be replaced. Welding a cracked crankshaft is not a proper repair since the crankshaft will fail rapidly in use.

Answer C is incorrect. Only Technician A is correct.

Answer D is incorrect. Technician A is correct.

39. Technician A says that the lobe face on a flat tappet lifter camshaft is flat. Technician B says a flat tappet lifter will have a concave face. Who is correct?

TASK B.11

A. A only

B. B only

C. Both A and B

D. Neither A nor B

Answer A is incorrect. A flat tappet camshaft will have a slight side-to-side taper to cause lifter spin and reduce wear.

Answer B is incorrect. A flat tappet lifter end will be slightly convex to assist in lifter rotation to reduce wear on the lifter and camshaft.

Answer C is incorrect. Neither Technician is correct.

Answer D is correct. Neither Technician is correct. The surface of a flat tappet camshaft is slightly tapered and the lifter will have a convex face.

40. Technician A says that oil galley plugs in the block may be pipe thread. Technician B says the oil galley plugs may be cup plugs. Who is correct?

TASK C.3

A. A only

B. B only

C. Both A and B

D. Neither A nor B

Answer A is incorrect. Technician B is also correct.

Answer B is incorrect. Technician A is also correct.

Answer C is correct. Both Technicians are correct. A block may have just pipe thread plugs in the oil galley or it might have pipe thread plugs and cup plugs.

Answer D is incorrect. Both Technicians are correct.

41. Technician A says a timing chain can be checked for stretch by using a timing light. Technician B says a stretched chain may cause intermittent poor performance. Who is correct?

TASK B.13

A. A only

B. B only

C. Both A and B

D. Neither A nor B

Answer A is incorrect. Technician B is also correct.

Answer B is incorrect. Technician A is also correct.

Answer C is correct. Both Technicians are correct. With the engine running, the timing mark can be watched while the engine throttle is snapped open and released. Excessive chain stretch will cause rapid fluctuation of the timing mark while RPM is coming down. A stretched chain may cause poor performance since the crankshaft-to-camshaft orientation is constantly changing.

Answer D is incorrect. Both Technicians are correct.

TASK D.9

42. A thermostat in the cooling system helps prevent:

    A. Overheating.

    B. Overcooling.

    C. Heater core leaks.

    D. Hose deterioration.

Answer A is incorrect. A thermostat's primary responsibility is to prevent overcooling.

Answer B is correct. A thermostat opens and closes to keep the engine at a normal temperature and prevents overcooling.

Answer C is incorrect. Heater core leaks are caused by corrosion.

Answer D is incorrect. Hose deterioration is caused by contaminants and age.

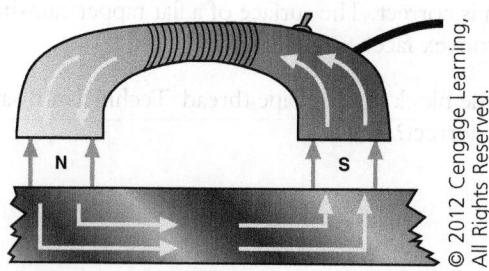

© 2012 Cengage Learning, All Rights Reserved.

TASK B.2

43. An electromagnetic-type tester shown in the figure above, along with iron filings, can be used to check for cracks in:

    A. Aluminum intake manifolds.

    B. Pistons.

    C. Aluminum cylinder heads.

    D. Cast iron cylinder heads.

Answer A is incorrect. Aluminum is not a ferrous metal so a magnet will not work for crack checks.

Answer B is incorrect. Pistons are an aluminum alloy and non-ferrous.

Answer C is incorrect. Aluminum is non-ferrous and therefore is not attracted to a magnet.

Answer D is correct. A cast iron head is ferrous and will conduct a magnetic field.

TASK B.10

44. The rocker arms on a pushrod engine have a 1.5:1 ratio. This means that:

    A. A cam lift of 0.350 inches will cause the valve to open 0.175 inches.

    B. A cam lift of 0.350 inches will cause the valve to open 0.525 inches.

    C. A cam lift of 0.350 inches will cause the valve to open 0.350 inches.

    D. A cam lift of 0.350 inches will cause the valve to open 0.700 inches.

Answer A is incorrect. The valve will open 0.525 inches.

Answer B is correct. A rocker arm ratio of 1.5:1 means the valve will move 1.5 times more than the pushrod, or 0.525 inches.

Answer C is incorrect. The valve will open 0.525 inches.

Answer D is incorrect. The valve will open 0.525 inches.

45. Accessory drive belts are being inspected. Technician A says the automatic tensioner should not vibrate when the engine is running without the A/C engaged. Technician B says all pulleys driven by a belt must be lined up. Who is correct?

    A. A only

    B. B only

    C. Both A and B

    D. Neither A nor B

TASK D.7

Answer A is incorrect. Technician B is also correct.

Answer B is incorrect. Technician A is also correct.

Answer C is correct. Both Technicians are correct. An automatic tensioner should not vibrate or jump when the engine is running without an A/C compressor engaged; if it does, it is loading and unloading the tension on the belt and will cause early failure. If the pulleys driven by a belt are not in line with each other, it will cause belt fraying and possibly cause the belt to jump off.

Answer D is incorrect. Both Technicians are correct.

46. The cylinders in a block have a pronounced ridge at the top. Technician A says this could be caused by excess fuel in the combustion chamber. Technician B says this could be caused by ring gap that was too tight. Who is correct?

    A. A only

    B. B only

    C. Both A and B

    D. Neither A nor B

TASK C.4

Answer A is correct. Only Technician A is correct. Excess fuel in the combustion chamber will wash the oil from the top of the cylinder wall and cause excessive wear.

Answer B is incorrect. Too tight a ring gap will cause the ring to bind in the cylinder when it warms up and cause ring breakage and piston damage.

Answer C is incorrect. Only Technician A is correct.

Answer D is incorrect. Technician A is correct.

47. Technician A says supercharger lubrication is done by the engine lubrication system. Technician B says supercharger oil is separate and should be changed according to manufacturer's specifications. Who is correct?

    A. A only

    B. B only

    C. Both A and B

    D. Neither A nor B

TASK E.3

Answer A is incorrect. Superchargers carry their own oil and are not lubricated with engine oil.

Answer B is correct. Only Technician B is correct. Superchargers carry their own oil and that oil should be changed according to manufacturer's specifications.

Answer C is incorrect. Only Technician B is correct.

Answer D is incorrect. Technician B is correct.

TASK C.14

48. Technician A says in-block camshaft bores do not warp so they will not have to be checked. Technician B says all in-block camshaft bearings are usually the same size. Who is correct?

A. A only

B. B only

C. Both A and B

D. Neither A nor B

Answer A is incorrect. In-block camshaft bores may warp and should be checked.

Answer B is incorrect. In-block camshaft bearings usually vary in diameter, larger in the front and slightly smaller through the bore. This helps keep the camshaft from walking front to rear.

Answer C is incorrect. Neither Technician is correct.

Answer D is correct. Neither Technician is correct. If the block has been subjected to extreme heat or stress, all bores should be checked for warp including the in-block camshaft bores. The camshaft bores are usually slightly different diameters to help control camshaft movement front-to-rear during engine operation.

TASK D.2

49. Technician A says the oil pump may be driven by the camshaft through the distributor. Technician B says the oil pump may be driven by the crankshaft. Who is correct?

A. A only

B. B only

C. Both A and B

D. Neither A nor B

Answer A is incorrect. Technician B is also correct.

Answer B is incorrect. Technician A is also correct.

Answer C is correct. Both Technicians are correct. The oil pump can be driven by the camshaft through the distributor or it can be driven by the end of the crankshaft.

Answer D is incorrect. Both Technicians are correct.

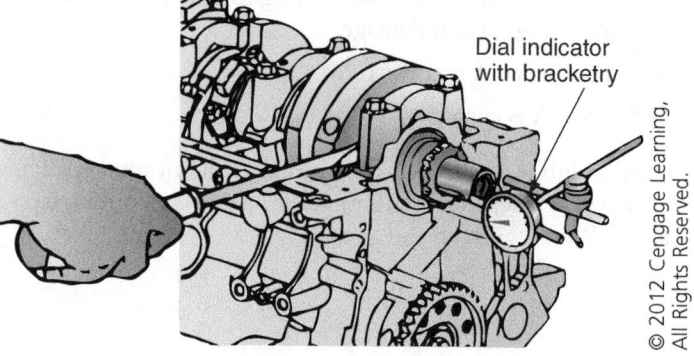

Dial indicator with bracketry

© 2012 Cengage Learning, All Rights Reserved.

TASK C.5

50. In the figure above, the technician is:

A. Measuring main bearing clearance.

B. Measuring rod bearing clearance.

C. Checking for crankshaft warp.

D. Checking thrust bearing clearance.

Answer A is incorrect. The Technician is checking thrust bearing clearance.

Answer B is incorrect. The Technician is checking thrust bearing clearance.

Answer C is incorrect. The Technician is checking thrust bearing clearance.

Answer D is correct. The Technician is checking thrust bearing clearance with a dial indicator and screwdriver.

PREPARATION EXAM 5—ANSWER KEY

1.	B	21.	C	41.	A
2.	D	22.	C	42.	B
3.	D	23.	C	43.	C
4.	C	24.	B	44.	C
5.	B	25.	D	45.	A
6.	A	26.	C	46.	B
7.	C	27.	A	47.	C
8.	D	28.	B	48.	C
9.	D	29.	D	49.	C
10.	B	30.	B	50.	D
11.	C	31.	A		
12.	B	32.	B		
13.	C	33.	C		
14.	C	34.	D		
15.	B	35.	C		
16.	B	36.	D		
17.	C	37.	C		
18.	C	38.	D		
19.	D	39.	A		
20.	B	40.	C		

PREPARATION EXAM 5—EXPLANATIONS

1. The customer complains of a loud noise coming from the engine compartment. Which of the following would be the LEAST LIKELY to cause the noise?

    TASK A.8

    A. Cracked exhaust manifold
    B. A vacuum leak
    C. A cracked flexplate
    D. Carbon buildup on the pistons

    Answer A is incorrect. A cracked exhaust manifold would cause a cracking or rumbling noise when cold.

    Answer B is correct. A vacuum leak would typically not cause a loud noise that the customer could hear from inside the car. A vacuum leak would be a high-pitched whistle.

    Answer C is incorrect. A cracked flexplate would create a loud snapping or knocking noise at idle which might diminish as speed increased.

    Answer D is incorrect. Carbon buildup on the pistons will cause a knocking noise from the affected cylinder.

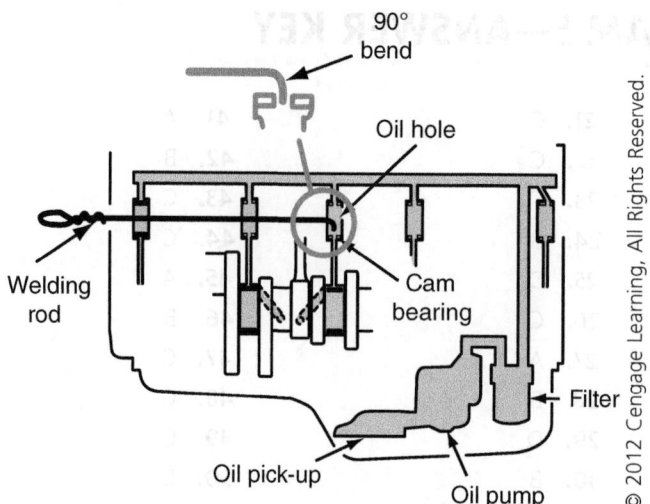

© 2012 Cengage Learning, All Rights Reserved.

TASK C.8

2. The figure above shows the technician:

    A. Installing in-block camshaft bearings.
    B. Cleaning the camshaft oil delivery holes.
    C. Removing in-block camshaft bearings.
    D. Checking journal oil hole and cam bearing oil hole alignment.

    Answer A is incorrect. Camshaft bearings are installed with a driver.

    Answer B is incorrect. The camshaft oil holes would be cleaned with a brush and air blow nozzle.

    Answer C is incorrect. The camshaft bearings are removed with a bearing driver tool.

    Answer D is correct. The camshaft bearings have been installed, and the technician is verifying alignment of oil supply hole and cam bearing hole.

TASK D.6

3. A high-mileage vehicle overheats only at highway speeds. Which of the following is the most likely to be the cause?

    A. A stuck closed thermostat
    B. A defective radiator cap vacuum valve
    C. An inoperative cooling fan
    D. A radiator whose core is clogged with deposits

    Answer A is incorrect. A stuck closed thermostat will cause overheating in city or low-speed driving.

    Answer B is incorrect. A radiator cap vacuum valve should only function during cooling off periods with the engine off.

    Answer C is incorrect. At highways speeds, the cooling fan function may be replaced by ram air effect.

    Answer D is correct. At highway speeds, the heat produced in the engine is greater. If the radiator cannot transfer that extra heat to the core because of blockage, it will cause engine overheating.

4. A battery rated at 600 cold cranking amps (CCA) is load tested at one-half of its rated CCA for 15 seconds. The results show 10.1 volts. This indicates the battery:

TASK E.4

   A. Is bad and should be replaced.
   B. Needs recharging.
   C. Is good.
   D. Should be retested for 30 seconds at load.

   Answer A is incorrect. The battery passed the load test.

   Answer B is incorrect. The battery passed the load test and should recover its lost voltage after a short period of time.

   Answer C is correct. The battery load test specification is one-half the cold cranking amps for 15 seconds, and the battery voltage should not drop below 9.6 volts at 70°F.

   Answer D is incorrect. No further testing is needed; the battery passed the load test.

5. Technician A says a full-floating piston pin will ride in a steel bushing in the small end of the connecting rod. Technician B says a bronze bushing is used. Who is correct?

TASK C.11

   A. A only
   B. B only
   C. Both A and B
   D. Neither A nor B

   Answer A is incorrect. A steel bushing is not used.

   Answer B is correct. Only Technician B is correct. A bronze bushing is used because of its good lubricating properties.

   Answer C is incorrect. Only Technician B is correct.

   Answer D is incorrect. Technician B is correct.

6. Technician A says that o-ring type valve stem seals are installed after the spring and retainer are installed. Technician B says that positive lock valve stem seals ride on the valve stem. Who is correct?

TASK B.6

   A. A only
   B. B only
   C. Both A and B
   D. Neither A nor B

   Answer A is correct. Only Technician A is correct. O-ring type valve stem seals seal the spring retainer to the valve stem and are installed after the spring and retainer are installed and the spring is compressed.

   Answer B is incorrect. Positive lock valve stem seals are press fit onto the valve guide and the valve rides up and down through it.

   Answer C is incorrect. Only Technician A is correct.

   Answer D is incorrect. Technician A is correct.

TASK A.5

7. A vehicle is brought in with excessive oil loss. Technician A says the loss may be through the rear main seal. Technician B says bad valve stem seals can cause excessive oil consumption. Who is correct?

    A. A only
    B. B only
    C. Both A and B
    D. Neither A nor B

    Answer A is incorrect. Technician B is also correct.

    Answer B is incorrect. Technician A is also correct.

    Answer C is correct. Both Technicians are correct. Oil can leak past a worn rear main seal. Bad valve stem seals will cause oil to be drawn into the combustion chamber and burned.

    Answer D is incorrect. Both Technicians are correct.

TASK C.12

8. Technician A says that piston ring gap that is too wide could cause ring binding and breakage. Technician B says that piston ring gap that is too tight would cause combustion gas blowby. Who is correct?

    A. A only
    B. B only
    C. Both A and B
    D. Neither A nor B

    Answer A is incorrect. Ring gap that is too wide will cause blowby.

    Answer B is incorrect. Ring gap that is too tight will cause binding and breakage.

    Answer C is incorrect. Neither Technician is correct.

    Answer D is correct. Neither Technician is correct. The rings should seal the cylinder as tightly as possible without binding. If the ring gap is too wide, the ring will not seal properly as it expands from heat and will have enough gap to allow excessive combustion gases (blowby) into the crankcase and result in a loss of power. A ring that has too little gap will become tight in the cylinder bore as it expands from heat causing binding on the cylinder wall, cylinder scuffing, and ring or piston damage.

TASK D.1

9. A customer brings his car to you complaining of a loss of oil pressure after extended driving. Which of the following is the most likely cause?

    A. Worn main bearings
    B. Weak oil pump
    C. Defective oil sending unit
    D. Trash in the oil pan stopping-up the oil pickup screen

    Answer A is incorrect. Worn main bearings will cause low oil pressure all the time.

    Answer B is incorrect. A weak oil pump will cause low oil pressure all the time.

    Answer C is incorrect. A defective oil sending unit will show low pressure all the time.

    Answer D is correct. If the complaint is loss of oil pressure after extended driving, suspect trash in the oil pan being picked up by the oil pickup screen; this will block oil flow to the pump.

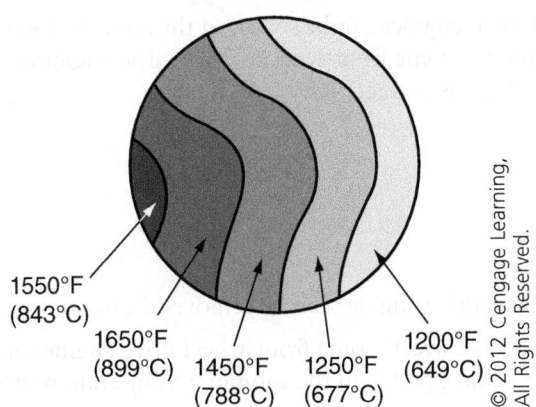

© 2012 Cengage Learning, All Rights Reserved.

1550°F
(843°C)

1650°F
(899°C) 1450°F   1250°F   1200°F
          (788°C)   (677°C) (649°C)

10. Technician A says the figure above shows the valve head temperatures for a properly seated valve. Technician B says the figure shows the valve head temperatures for a valve that is not properly contacting the seat. Who is correct?

TASK B.8

A. A only

B. B only

C. Both A and B

D. Neither A nor B

Answer A is incorrect. The head temperature for properly seated valves will form concentric circles, with the center being the hottest.

Answer B is correct. Only Technician B is correct. An improperly seated valve will be hotter on the outside at the point of poor contact with the seat.

Answer C is incorrect. Only Technician B is correct.

Answer D is incorrect. Technician B is correct.

11. An engine cranks over very slowly and may not start. Which is the LEAST LIKELY cause?

TASK A.2

A. Excessive electrical resistance at the starter

B. Weak battery

C. Broken timing belt

D. Worn ignition switch contacts

Answer A is incorrect. Excessive electrical resistance may reduce the current flow to the starter causing slow cranking speed.

Answer B is incorrect. A weak battery may not supply enough current to spin the starter rapidly.

Answer C is correct. A broken timing belt will cause rapid crank speeds since no compression is being developed and causing a no-start condition.

Answer D is incorrect. Worn ignition switch contacts may prevent no, or poor, current flow to the starting circuit.

TASK C.1

12. Technician A says that all front-wheel drive engines can be removed through the top of the engine compartment. Technician B says that some front-wheel drive vehicle engines must be removed from underneath the vehicle. Who is correct?

    A.   A only

    B.   B only

    C.   Both A and B

    D.   Neither A nor B

Answer A is incorrect. Not all front-wheel drive engines can be removed from the top.

Answer B is correct. Only Technician B is correct. Some front-wheel drive engine removals require the engine and transmission to be dropped with the subframe from underneath the vehicle.

Answer C is incorrect. Only Technician B is correct.

Answer D is incorrect. Technician B is correct.

TASK A.3

13. A customer's vehicle is leaking oil on his driveway from the rear of the engine. Technician A says the leak may be coming from the rear main seal. Technician B says if a visual inspection does not pinpoint the leak, florescent dye should be put in the oil. Who is correct?

    A.   A only

    B.   B only

    C.   Both A and B

    D.   Neither A nor B

Answer A is incorrect. Technician B is also correct.

Answer B is incorrect. Technician A is also correct.

Answer C is correct. Both Technicians are correct. An oil leak at the rear of the engine may be the rear main seal but if visual inspection is inconclusive, dye should be put in the oil, the engine run, and then check for leaks with a blacklight.

Answer D is incorrect. Both Technicians are correct.

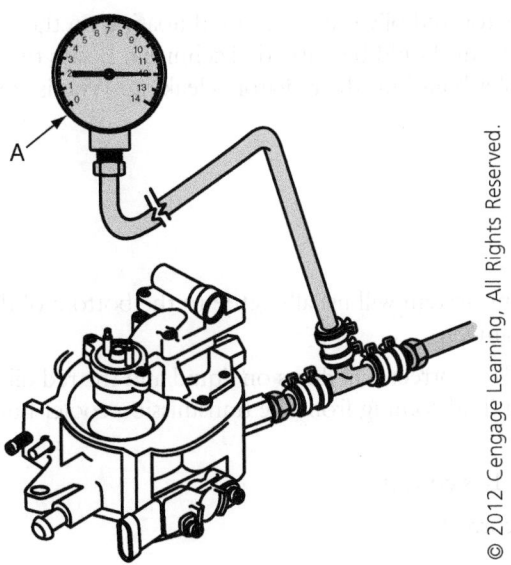

© 2012 Cengage Learning, All Rights Reserved.

14. In the figure above, the technician is checking:

A. Oil pressure.

B. Engine vacuum.

C. Fuel pressure.

D. The PCV system.

TASK E.1

Answer A is incorrect. Oil pressure would be checked at the oil sending unit.

Answer B is incorrect. Engine vacuum is checked with a vacuum gauge, not a pressure gauge.

Answer C is correct. The technician is checking fuel pressure at the throttle body.

Answer D is incorrect. If the PCV system were being checked, it would be with a vacuum gauge.

15. During a vacuum check of a poorly running engine, it is found that vacuum holds steady between 8 and 14 in. Hg. Technician A says this may be caused by weak valve springs. Technician B says ignition timing that is off could cause this problem. Who is correct?

A. A only

B. B only

C. Both A and B

D. Neither A nor B

TASK A.6

Answer A is incorrect. Weak valve springs would result in erratic needle movement.

Answer B is correct. Only Technician B is correct. Ignition timing that is off may cause low, steady vacuum readings due to improper spark timing.

Answer C is incorrect. Only Technician B is correct.

Answer D is incorrect. Technician B is correct.

TASK D.6

16. While checking coolant level in the radiator, red oily specs are noted floating on the coolant. Technician A says this is rust and the system should be flushed. Technician B says this is probably transmission fluid and the cooler built into the radiator is leaking. Who is correct?

   A. A only

   B. B only

   C. Both A and B

   D. Neither A nor B

   Answer A is incorrect. Rust in the cooling system will usually settle to the bottom of the radiator since it is heavier than the coolant.

   Answer B is correct. Only Technician B is correct. Oil floats on liquid and the red oily specs in the coolant are probably transmission fluid coming from the transmission cooler built into the radiator.

   Answer C is incorrect. Only Technician B is correct.

   Answer D is incorrect. Technician B is correct.

TASK A.9

17. The technician has done a compression test on an engine and found cylinders #1 and #3 to have low compression. What would be the next test he should do in his diagnosis?

   A. Power balance test

   B. Cylinder leakage test

   C. Wet compression test

   D. Dynamic compression test

   Answer A is incorrect. A power balance test would not be done. He is looking for a mechanical failure, not an ignition or fuel failure.

   Answer B is incorrect. A cylinder leakage test may need to be done but it is not the next test.

   Answer C is correct. A wet compression test will indicate whether the loss of compression is through the rings or somewhere else.

   Answer D is incorrect. A dynamic compression test checks for proper breathing in an operating engine and would not be the next test.

TASK B.11

18. Technician A says a hydraulic lifter in a pushrod engine may pump up during very high RPM operation. Technician B says this may cause bent valves. Who is correct?

   A. A only

   B. B only

   C. Both A and B

   D. Neither A nor B

   Answer A is incorrect. Technician B is also correct.

   Answer B is incorrect. Technician A is also correct.

   Answer C is correct. Both Technicians are correct. During high RPM operation in a pushrod engine, the lifter may pump up and hold the valve open due to slight amounts of lash in the valve train. When this happens, the valve may be held open and hit by the piston; at the very least, the engine will stall.

   Answer D is incorrect. Both Technicians are correct.

19. The technician test drives the customer's car but cannot duplicate the customer's complaint. His next step should be:

    A. Ask the service writer to call the customer for more information.

    B. Return the vehicle to the customer with the comment, "No problem found."

    C. Do a repair based on the customer's complaint.

    D. Call the customer to get more information such as when and how often the problem occurs.

TASK A.1

Answer A is incorrect. The service writer talked to the customer when he brought the vehicle in and may not know the right questions to ask.

Answer B is incorrect. The vehicle was brought in for a problem and the customer expects a repair, although sometimes an intermittent problem cannot be duplicated.

Answer C is incorrect. No repairs should be made if the complaint cannot be verified.

Answer D is correct. If the technician calls the customer, he can ask the pertinent questions to help identify the problem.

20. Technician A says all piston pins are centered in the piston skirts. Technician B says some piston pins might be offset. Who is correct?

    A. A only

    B. B only

    C. Both A and B

    D. Neither A nor B

TASK C.10

Answer A is incorrect. Some piston pins are centered, others are offset from center.

Answer B is correct. Only Technician B is correct. Some piston pins on V-engines are offset toward the major thrust surface to reduce piston rock under combustion pressures.

Answer C is incorrect. Only Technician B is correct.

Answer D is incorrect. Technician B is correct.

21. A cranking compression test on a 4-cylinder engine has been done and the results were:

CYL #1	CYL #2	CYL #3	CYL #4
135 psi	20 psi	20 psi	130 psi

    Technician A says a wet test should be done on cylinders #2 and #3. Technician B says the results indicate a blown head gasket between cylinders #2 and #3. Who is correct?

    A. A only

    B. B only

    C. Both A and B

    D. Neither A nor B

TASK A.8

Answer A is incorrect. Technician B is also correct.

Answer B is incorrect. Technician A is also correct.

Answer C is correct. Both Technicians are correct. A wet test will enable the technician to eliminate rings as a possible problem. Typically, the same low compression readings on adjacent cylinders indicate a blown head gasket.

Answer D is incorrect. Both Technicians are correct.

TASK D.7

22. A serpentine belt has just been replaced; it squeals in the morning and during acceleration. Technician A says this could be caused by worn pulley grooves. Technician B says a weak tensioner could cause the problem. Who is correct?

    A.   A only.

    B.   B only.

    C.   Both A and B.

    D.   Neither A nor B.

Answer A is incorrect. Technician B is also correct.

Answer B is incorrect. Technician A is also correct.

Answer C is correct. Both Technicians are correct. Worn pulley grooves could cause the belt to slip under load and if the tensioner is not supplying enough tension during start up, the belt will squeal as the alternator charges at higher outputs.

Answer D is incorrect. Both Technicians are correct.

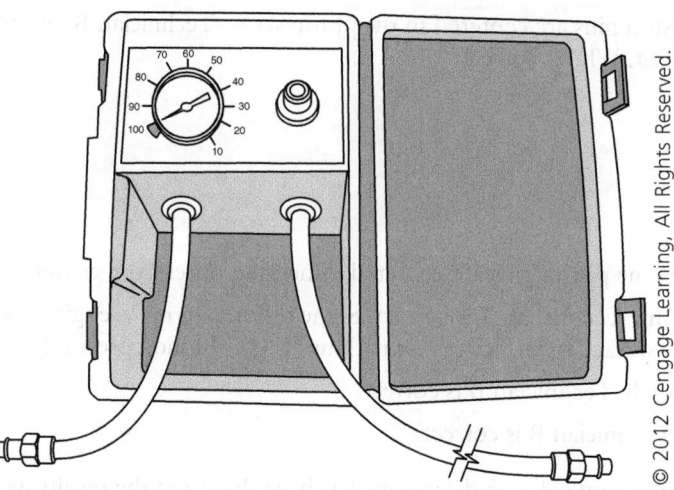

© 2012 Cengage Learning, All Rights Reserved.

TASK A.9

23. The gauge set shown in the figure above is used to perform:

    A.   Oil pressure tests

    B.   Compression tests

    C.   Cylinder leakage tests

    D.   Vacuum tests

Answer A is incorrect. An oil pressure gauge set only has one gauge.

Answer B is incorrect. A compression tester will only have one gauge.

Answer C is correct. This is a cylinder leakage tester. One gauge reads psi input to the cylinder; the second gauge registers the psi the cylinder is holding.

Answer D is incorrect. A vacuum gauge set will only have one gauge.

24. Technician A says RTV can be used to seal a throttle body injection assembly to the intake manifold. Technician B says RTV is a gasket maker and should not be applied to another gasket. Who is correct?

    A. A only

    B. B only

    C. Both A and B

    D. Neither A nor B

TASK C.16

Answer A is incorrect. RTV should not be used in applications where fuel can come in contact with it. Fuel will cause deterioration of the seal.

Answer B is correct. Only Technician B is correct. RTV is made to be used by itself as a gasket.

Answer C is incorrect. Only Technician B is correct.

Answer D is incorrect. Technician B is correct.

25. The exhaust on a customer's car is white and has a sweet smell. Technician A says this indicates a rich condition in the cylinder. Technician B says this indicates oil being burned in the cylinder. Who is correct?

    A. A only

    B. B only

    C. Both A and B

    D. Neither A nor B

TASK A.5

Answer A is incorrect. An engine that is running too rich will have black exhaust.

Answer B is incorrect. Oil burning in the cylinders will cause the exhaust to be bluish.

Answer C is incorrect. Neither Technician is correct.

Answer D is correct. Neither Technician is correct. White color exhaust accompanied by a sweet smell indicates coolant is entering the cylinder.

26. Technician A says most newly designed engines use low tension piston rings. Technician B says that low tension piston rings decrease friction in the cylinder and help improve fuel economy. Who is correct?

    A. A only

    B. B only

    C. Both A and B

    D. Neither A nor B

TASK C.12

Answer A is incorrect. Technician B is also correct.

Answer B is incorrect. Technician A is also correct.

Answer C is correct. Both Technicians are correct. Most newly designed engines use low tension piston rings to reduce friction and improve fuel economy.

Answer D is incorrect. Both Technicians are correct.

TASK A.2

27. Which of these should be performed first when a starter fails to crank the engine?

    A. Measure static battery voltage.
    B. Remove and check spark plugs.
    C. Check for fuel pressure.
    D. By-pass the starter solenoid with a remote starter button.

    Answer A is correct. The battery voltage should be checked first to verify there is enough power to turn the starter.

    Answer B is incorrect. Spark plugs may be removed if a hydro-lock condition is suspected in the cylinder. This would not be the first test.

    Answer C is incorrect. Fuel pressure that is not at specifications may cause an engine to not start but will not keep it from cranking.

    Answer D is incorrect. The correct battery voltage should be checked first. The solenoid can be by-passed if there is a faulty ignition switch or neutral safety switch.

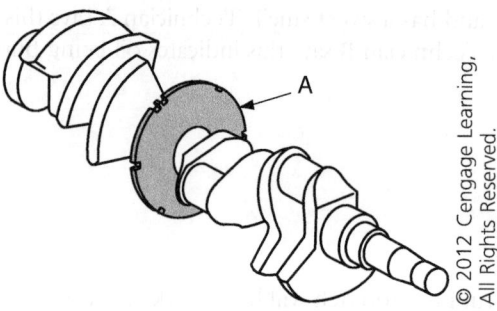

A

© 2012 Cengage Learning, All Rights Reserved.

TASK E.5

28. In the figure above, Technician A says that A is a counter balance weight. Technician B says it is a reluctor ring. Who is correct?

    A. A only
    B. B only
    C. Both A and B
    D. Neither A nor B

    Answer A is incorrect. A is the reluctor ring, which is used with the crankshaft position sensor to tell the ECM where the pistons are.

    Answer B is correct. Only Technician B is correct. A is the reluctor ring, which is used with the crankshaft position sensor to tell the ECM where the pistons are.

    Answer C is incorrect. Only Technician B is correct.

    Answer D is incorrect. Technician B is correct.

29. All of the following are causes of low engine oil pressure EXCEPT:

    A. Worn crankshaft main bearings.

    B. Weak oil pump pressure relief valve spring.

    C. Stuck open pressure relief valve.

    D. Restricted pushrod oil passages.

TASK D.2

Answer A is incorrect. Worn crankshaft main bearings will cause low oil pressure.

Answer B is incorrect. A weak oil pump pressure relief valve spring will not allow oil pressure to build to the proper pressure before allowing the oil to return to the oil pan from the oil pump.

Answer C is incorrect. A stuck open pressure relief valve will cause low oil pressure.

Answer D is correct. Restricted pushrod oil passages will affect oil delivery to the rocker arms, but will not affect oil pressure.

30. Technician A says a thermostat rated 180 degrees is fully open at that temperature. Technician B says the thermostat begins to open at 180 degrees. Who is correct?

    A. A only

    B. B only

    C. Both A and B

    D. Neither A nor B

TASK D.9

Answer A is incorrect. A thermostat rated at 180 degrees begins to open at that temperature.

Answer B is correct. Only Technician B is correct. A thermostat rated at 180 degrees begins to open at that temperature and is fully open at about 220 degrees.

Answer C is incorrect. Only Technician B is correct.

Answer D is incorrect. Technician B is correct.

31. Technician A says if a timing belt is removed and is going to be reused, the direction of rotation must be marked. Technician B says a timing belt that has been contaminated with oil can be cleaned and reused. Who is correct?

    A. A only

    B. B only

    C. Both A and B

    D. Neither A nor B

TASK B.13

Answer A is correct. Only Technician A is correct. A used timing belt that is removed and will be reused must be put back on in the same direction of rotation it came off; failure to do this may cause belt failure.

Answer B is incorrect. A timing belt that has been contaminated with oil will deteriorate and fail. If it is cleaned, the oil that has saturated the belt will not be removed.

Answer C is incorrect. Only Technician A is correct.

Answer D is incorrect. Technician A is correct.

TASK D.10

32. Technician A says old coolant is not hazardous and can be disposed of in the shop drain. Technician B says old coolant should be disposed of in an environmentally safe manner. Who is correct?

A. A only

B. B only

C. Both A and B

D. Neither A nor B

Answer A is incorrect. Old coolant is considered hazardous waste in most states, and it is illegal to dispose of in a floor drain.

Answer B is correct. Only Technician B is correct. Old coolant should be picked up by a licensed waste disposal company and disposed of in an environmentally safe manner.

Answer C is incorrect. Only Technician B is correct.

Answer D is incorrect. Technician B is correct.

TASK C.5

33. When inspected, the crankshaft bearings that exhibit more wear on the bearings the furthest from the oil pump, which is less noticeable as you get closer to the oil pump. What is the most likely reason for the wear?

A. Engine overloading at low RPM

B. Over-tightened accessory belts

C. Dry starts

D. A warped crankshaft

Answer A is incorrect. Engine overloading would cause wear on all the lower main bearings.

Answer B is incorrect. Wear from over-tightened accessory belts would start at the top of main bearing #1 and transfer to the lower bearing on the last journal.

Answer C is correct. A condition known as dry starts occurs when the engine has not been operated in a while. If there is no oil on the main bearing surfaces and during start up, the last journal to get oil pressure will show more wear.

Answer D is incorrect. A warped crankshaft will usually show up as excessive wear on one or two of the center main bearings.

TASK A.4

34. An engine idles rough and the technician notes a high-pitch whistle from the engine compartment. Which of the following would most likely be the cause?

A. A slipping fan belt

B. A leaking exhaust pipe

C. A defective alternator bearing

D. A vacuum leak on the intake

Answer A is incorrect. A slipping fan belt will squeal, not whistle.

Answer B is incorrect. A leaking exhaust manifold will make a popping noise or rumbling.

Answer C is incorrect. A defective alternator bearing will growl.

Answer D is correct. A vacuum leak on the intake can cause a rough idle if it is after the airflow meter. Due to unmetered air entering the engine, it will cause a high-pitch whistle as the air is drawn into the engine.

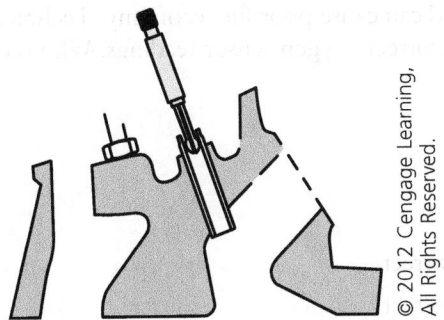

© 2012 Cengage Learning, All Rights Reserved.

35. The figure above shows a technician:

    A. Installing a valve guide.
    B. Measuring stem installed height.
    C. Measuring a valve guide for wear.
    D. Removing a valve guide.

    TASK B.7

    Answer A is incorrect. The technician is measuring the valve guide.

    Answer B is incorrect. The technician is measuring the valve guide.

    Answer C is correct. The technician is measuring the valve guide for wear to determine necessary repairs.

    Answer D is incorrect. The technician is measuring the valve guide.

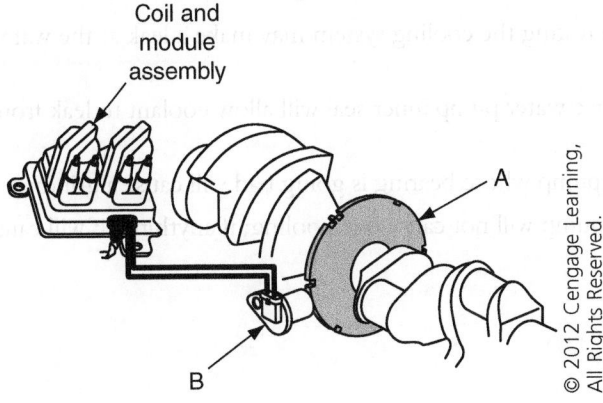

Coil and module assembly

A

B

© 2012 Cengage Learning, All Rights Reserved.

36. In the figure above, Technician A says if the distance between A and B is too wide, it will cause a no-crank condition. Technician B says pickup B is a thermistor. Who is correct?

    A. A only
    B. B only
    C. Both A and B
    D. Neither A nor B

    TASK E.6

    Answer A is incorrect. Too large a gap between the crankshaft reluctor ring and the crankshaft position sensor will cause no-trigger signal and a no-start condition; it will not affect cranking.

    Answer B is incorrect. The crankshaft position sensor is a permanent magnet-type sensor, not a thermistor.

    Answer C is incorrect. Neither Technician is correct.

    Answer D is correct. Neither Technician is correct. The reluctor ring on the crankshaft has gaps in it that correspond with piston position. The crankshaft position sensor emits a magnetic field that is interrupted when the gap passes it, letting the PCM know the position of that cylinder to fire the sparkplug at the proper time. If the air gap between the sensor and the reluctor ring is too wide, the PCM will receive a weak or no signal and will not fire the sparkplug.

TASK E.7

37. Technician A says a cracked exhaust manifold can cause poor fuel economy. Technician B says a cracked exhaust manifold will cause incorrect oxygen sensor readings. Who is correct?

A. A only

B. B only

C. Both A and B

D. Neither A nor B

Answer A is incorrect. Technician B is also correct.

Answer B is incorrect. Technician A is also correct.

Answer C is correct. Both Technicians are correct. A cracked exhaust manifold will allow air to be drawn into the exhaust and cause the oxygen sensor to sense a lean condition, which will cause the ECM to increase fuel delivery resulting is decreased fuel economy.

Answer D is incorrect. Both Technicians are correct.

TASK D.11

38. A defective water pump can be diagnosed by all of the following EXCEPT:

A. Using a pressure tester.

B. A coolant leak from the water pump.

C. A grinding noise from the pump area.

D. A lower than normal reading on the temperature gauge.

Answer A is incorrect. Pressure testing the cooling system may make a leak at the water pump more pronounced.

Answer B is incorrect. A defective water pump inner seal will allow coolant to leak from the water pump drain hole.

Answer C is incorrect. A water pump whose bearing is going bad will cause a grinding noise.

Answer D is correct. A water pump will not cause overcooling; if anything, if will cause overheating.

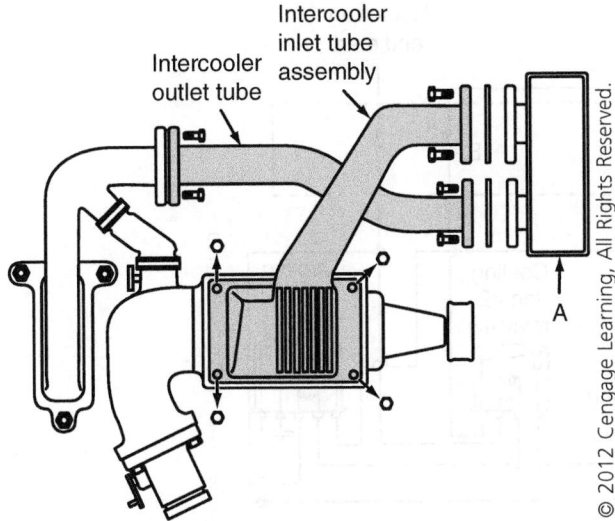

Intercooler inlet tube assembly

Intercooler outlet tube

A

© 2012 Cengage Learning, All Rights Reserved.

39. In the figure above, Technician A says that B is the intercooler. Technician B says a supercharger is driven by exhaust gases. Who is correct?

TASK E.3

A. A only

B. B only

C. Both A and B

D. Neither A nor B

Answer A is correct. Only Technician A is correct. B indicates the supercharger intercooler.

Answer B is incorrect. A supercharger is belt driven by the engine.

Answer C is incorrect. Only Technician A is correct.

Answer D is incorrect. Technician A is correct.

40. All of the following are true about TTY bolts EXCEPT:

TASK B.17

A. Provide a more uniform clamping force.

B. Usually require special tightening procedures.

C. Must always be discarded after use.

D. May be used in connecting rod applications.

Answer A is incorrect. TTY bolts provide more uniform clamping force.

Answer B is incorrect. TTY bolts are tightened to a specified torque, then turned an additional number of degrees.

Answer C is correct. TTY bolts may be reused after being checked for stretch following manufacturer's recommendations.

Answer D is incorrect. TTY bolts are used in some connecting rod assemblies.

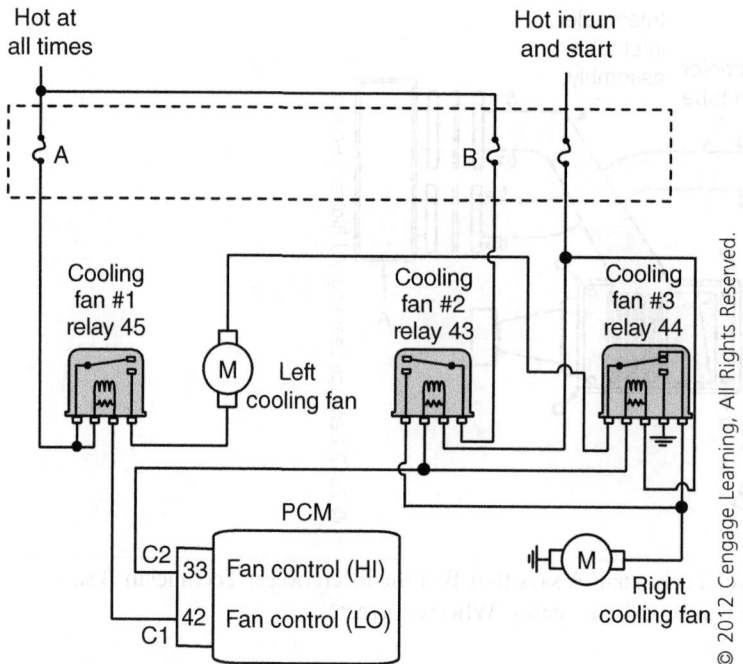

Hot at all times

Hot in run and start

A

B

Cooling fan #1 relay 45

Cooling fan #2 relay 43

Cooling fan #3 relay 44

M Left cooling fan

© 2012 Cengage Learning, All Rights Reserved.

PCM

C2

33 Fan control (HI)

42 Fan control (LO)

C1

M Right cooling fan

TASK D.12

41. In the figure above, Technician A says a blown fuse at point A will keep the left cooling fan from operating. Technician B says a blown fuse at point B will keep the left cooling fan from running in low speed. Who is correct?

   A. A only

   B. B only

   C. Both A and B

   D. Neither A nor B

   Answer A is correct. Only Technician A is correct. Fuse A supplies current for the apply side of right cooling fan; if this fuse is blown, the right cooling fan will not operate.

   Answer B is incorrect. Fuse B supplies current for the apply side of the right cooling fan so fuse B will not affect the left cooling fan.

   Answer C is incorrect. Only Technician A is correct.

   Answer D is incorrect. Technician A is correct.

42. Technician A says that fuel injectors are controlled by the crankshaft position sensor. Technician B says a noid light can be used to check for injector trigger signal. Who is correct?

TASK E.1

   A.   A only

   B.   B only

   C.   Both A and B

   D.   Neither A nor B

Answer A is incorrect. Fuel injector pulse width is controlled by the ECM, which receives a signal from the crankshaft position sensor.

Answer B is correct. Only Technician B is correct. A noid light, plugged into the injector pigtail, is used to verify injector trigger signal.

Answer C is incorrect. Only Technician B is correct.

Answer D is incorrect. Technician B is correct.

43. Technician A says that valve keepers must be inspected for wear, cracks, and rounded corners. Technician B says that valve keeper lock grooves must also be checked for wear. Who is correct?

TASK B.5

   A.   A only

   B.   B only

   C.   Both A and B

   D.   Neither A nor B

Answer A is incorrect. Technician B is also correct.

Answer B is incorrect. Technician A is also correct.

Answer C is correct. Both Technicians are correct. Both the valve keepers and the valve keeper lock grooves in the valve stem must be checked for damage. If the keepers and grooves are not checked, the valve may be released while the engine is running causing damage to the cylinder and head.

Answer D is incorrect. Both Technicians are correct.

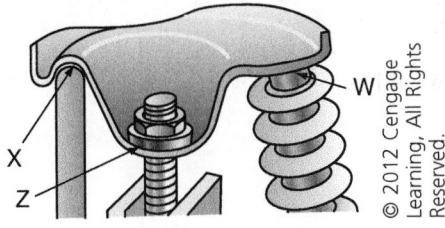

44. In the figure above, how is valve lash adjusted?

TASK B.12

   A.   By adding shims at point W

   B.   By adding shims at point X

   C.   By rotating nut Z

   D.   By replacing the pushrod

Answer A is incorrect. Shims are not added to adjust lash.

Answer B is incorrect. Shims are not added to adjust lash.

Answer C is correct. Lash is adjusted by turning nut Z.

Answer D is incorrect. The lash is adjusted by turning nut Z.

TASK E.2

45. Technician A says some manufacturers use a sensor to indicate air filter condition. Technician B says air filter condition is displayed on the dash. Who is correct?

    A.   A only

    B.   B only

    C.   Both A and B

    D.   Neither A nor B

    Answer A is correct. Only Technician A is correct. Some manufacturers us an air filter condition sensor in the air intake ductwork. As airflow is reduced by a dirty air filter, the sensor changes from green to red.

    Answer B is incorrect. The air filter sensor is not an electrical input on the dash display.

    Answer C is incorrect. Only Technician A is correct.

    Answer D is incorrect. Technician A is correct.

TASK B.14

46. The camshaft bore in an overhead cam engine does not use bearing inserts. Technician A says if the camshaft-to-bore clearance exceeds specifications, bearing inserts could be installed to correct the problem. Technician B says the head must be replaced. Who is correct?

    A.   A only

    B.   B only

    C.   Both A and B

    D.   Neither A nor B

    Answer A is incorrect. Most heads that do not use camshaft bearing inserts do not have enough clearance in the bore to additionally bore the head oversize and install bearing inserts.

    Answer B is correct. Only Technician B is correct. If camshaft-to-bore clearance exceeds specifications, the head must be replaced.

    Answer C is incorrect. Only Technician B is correct.

    Answer D is incorrect. Technician B is correct.

TASK B.15

47. The LEAST LIKELY cause of camshaft bind would be:

    A.   Improperly installed bearings.

    B.   Bore misalignment.

    C.   Excessive bearing clearance.

    D.   Mixed up journal caps.

    Answer A is incorrect. Improperly installed bearings could cause camshaft bind.

    Answer B is incorrect. Bore misalignment and warp will cause camshaft bind.

    Answer C is correct. Excessive bearing clearance would cause oil runoff since the clearances are larger than specified, but they would not cause camshaft bind.

    Answer D is incorrect. Camshaft journal caps that are installed on the wrong journal can cause camshaft bind.

48. During a power balance test on a port fuel-injected engine, one cylinder is found to have virtually no RPM change. Which of these is the most likely cause?

    A. A faulty crankshaft position sensor
    B. A vacuum leak at the throttle body
    C. A defective plug wire
    D. A faulty camshaft position sensor

TASK A.7

Answer A is incorrect. A faulty crankshaft position sensor will affect all cylinders rather than just one.

Answer B is incorrect. A vacuum leak at the throttle body will affect all cylinders.

Answer C is correct. If the plug wire on that cylinder is defective, the cylinder will not fire and contribute to power.

Answer D is incorrect. A faulty camshaft position sensor would affect all cylinders.

49. Technician A says it is important to relieve pressure on an injector rail before replacing an injector. Technician B says fuel pressure may be 50 psi or higher, depending on the injection system. Who is correct?

    A. A only
    B. B only
    C. Both A and B
    D. Neither A nor B

TASK E.1

Answer A is incorrect. Technician B is also correct.

Answer B is incorrect. Technician A is also correct.

Answer C is correct. Both Technicians are correct. Anytime a fuel injection system is opened, it is important to relieve fuel pressure to prevent fuel spray and possible injury. The fuel system may have pressure greater that 50 psi, depending on the type of system.

Answer D is incorrect. Both Technicians are correct.

50. After a cold start, the customer notes a loud cracking or popping noise from the engine that quiets as the engine warms up. Technician A says this may be caused by a worn rod bearing. Technician B says this may be normal engine noise caused by lifter bleed down. Who is correct?

    A. A only
    B. B only
    C. Both A and B
    D. Neither A nor B

TASK A.4

Answer A is incorrect. A worn rod bearing will make a deep knocking noise, not cracking or popping, and the noise would not go away.

Answer B is incorrect. Lifter noise on cold starts is a light rattle noise.

Answer C is incorrect. Neither Technician is correct.

Answer D is correct. Neither Technician is correct. The noise is most likely caused by a cracked exhaust manifold or manifold gasket. The noise may go away after warm up as the metal expands.

PREPARATION EXAM 6—ANSWER KEY

1.	D	21.	B	41.	C
2.	C	22.	C	42.	C
3.	A	23.	B	43.	C
4.	D	24.	D	44.	B
5.	C	25.	B	45.	C
6.	D	26.	B	46.	C
7.	C	27.	C	47.	C
8.	C	28.	C	48.	A
9.	B	29.	A	49.	C
10.	C	30.	C	50.	C
11.	D	31.	A		
12.	D	32.	C		
13.	C	33.	B		
14.	C	34.	B		
15.	B	35.	C		
16.	D	36.	C		
17.	B	37.	C		
18.	C	38.	B		
19.	C	39.	B		
20.	B	40.	D		

PREPARATION EXAM 6—EXPLANATIONS

TASK C.1

1. Technician A says when removing an engine with an automatic transmission, the transmission torque converter is removed with the engine. Technician B says all the accessories must be removed from the engine before removal. Who is correct?

    A.   A only
    B.   B only
    C.   Both A and B
    D.   Neither A nor B

Answer A is incorrect. The transmission torque converter must be unbolted and pushed back into the transmission pump to prevent damage to the pump during engine removal.

Answer B is incorrect. It is usually possible to leave the alternator mounted on the engine during removal. It is best to take as little as possible from the engine before removal. After removing the engine, it is much easier to remove the parts.

Answer C is incorrect. Neither Technician is correct.

Answer D is correct. Neither Technician is correct. If an engine is removed with the torque converter still attached, the converter nose may bind in the transmission pump and break the pump housing. Due to tight engine compartments, any accessories, wiring, or other components that can be left on the engine during removal will be easier to remove with the engine out of the engine compartment and mounted on an engine stand. The components can be reinstalled before installation.

2. During teardown, it is noted that several valve stem tips were severely mushroomed. Technician A says the mushroomed tips must be dressed with a file before valve removal. Technician B says this was probably caused by the valve clearance being excessive. Who is correct?

TASK B.8

    A.   A only

    B.   B only

    C.   Both A and B

    D.   Neither A nor B

Answer A is incorrect. Technician B is also correct.

Answer B is incorrect. Technician A is also correct.

Answer C is correct. Both Technicians are correct. The normal cause for valve tip mushrooming is excessive clearance, which allows the rocker arm to hammer the valve tip rather than being in constant contact. If the tips are mushroomed, they cannot be removed without filing the tip. If they are driven out with a punch and hammer, the guides and valves will be damaged.

Answer D is incorrect. Both Technicians are correct.

3. Technician A says an intercooler may be used to cool the air taken in by the supercharger before it reaches the combustion chamber. Technician B says it is to prevent the supercharger from overheating. Who is correct?

TASK E.3

    A.   A only

    B.   B only

    C.   Both A and B

    D.   Neither A nor B

Answer A is correct. The intercooler reduces the heat content of the air from the supercharger before it enters the combustion chamber.

Answer B is incorrect. The intercooler does not cool the supercharger.

Answer C is incorrect. Only Technician A is correct.

Answer D is incorrect. Technician A is correct.

4. The technician is investigating an intermittent popping noise from under the hood that occurs when the RPM is raised. Which of the following is the LEAST LIKELY cause for the noise?

TASK A.4

    A.   Sticky intake valve

    B.   Incorrect ignition timing

    C.   Broken intake valve spring

    D.   Worn rod bearing

Answer A is incorrect. A sticky intake valve that is slow to close may be open enough to allow combustion to escape before it closes all the way.

Answer B is incorrect. If ignition timing is off, the spark may occur before the intake valve is completely closed allowing combustion to escape.

Answer C is incorrect. A broken or weak intake valve spring will not close the intake valve rapidly enough, or not at all, allowing combustion to escape.

Answer D is correct. A worn rod bearing noise would be constant and increase in frequency as the RPM is raised.

TASK B.10

5. Technician A says pushrods may be used to deliver oil to the rocker arms. Technician B says a bent pushrod may indicate a valve was struck by a piston. Who is correct?

A. A only

B. B only

C. Both A and B

D. Neither A nor B

Answer A is incorrect. Technician B is also correct.

Answer B is incorrect. Technician A is also correct.

Answer C is correct. Both Technicians are correct. A hollow pushrod is used to deliver lubrication to the rocker arms from the lifter. A pushrod may be the weakest link in the valve train, so if a piston hits a valve, the pushrod may bend before the valve stem.

Answer D is incorrect. Both Technicians are correct.

TASK A.5

6. The customer questions the large amount of blue smoke coming out of his tailpipe continuously. Technician A says the problem may be worn main bearings. Technician B says this may be caused by excess fuel in the combustion chamber. Who is correct?

A. A only

B. B only

C. Both A and B

D. Neither A nor B

Answer A is incorrect. Worn main bearings will make noise but will not allow oil to enter the combustion chamber.

Answer B is incorrect. Excessive fuel in the combustion chamber will produce black smoke from the tailpipe.

Answer C is incorrect. Neither Technician is correct.

Answer D is correct. Neither Technician is correct. Although worn main bearing may cause noise or engine failure, they will not cause oil to enter the combustion chamber; the rings will prevent that from occurring. Blue smoke is an indication of oil being burned in the combustion chamber. Excess unburned fuel in the combustion chamber is indicated by black smoke from the tailpipe.

TASK D.11

7. The water pump is being replaced on a rear-wheel drive vehicle with a longitudinally placed engine. What is the LEAST LIKELY part that will have to be removed?

A. The cooling fan

B. The fan belt

C. The radiator

D. The fan shroud

Answer A is incorrect. The cooling fan would be removed.

Answer B is incorrect. The belt turning the fan and the water pump would be removed.

Answer C is correct. The radiator would not have to be removed.

Answer D is incorrect. The fan shroud may have to be removed when the cooling fan is removed.

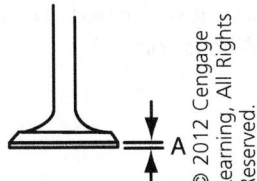

8. Technician A says that A in the figure above indicates the valve margin. Technician B says typical valve margin should measure no less than 0.031. Who is correct?

    A. A only
    B. B only
    C. Both A and B
    D. Neither A nor B

    TASK B.8

    Answer A is incorrect. Technician B is also correct.

    Answer B is incorrect. Technician A is also correct.

    Answer C is correct. Both Technicians are correct. The figure shows the valve margin. Typical valve margin after machining should be no less than 0.031. This may vary slightly for different engines; always check the manufacturer's specifications.

    Answer D is incorrect. Both Technicians are correct.

9. A poor performing engine has a cranking compression test done. The test fails with all cylinders showing low compression. The technician did a wet compression test and all cylinders increased substantially. Technician A says the wet compression test results indicate the head gasket is not sealing. Technician B says the wet compression test results indicate cylinder and ring wear. Who is correct?

    TASK A.8

    A. A only
    B. B only
    C. Both A and B
    D. Neither A nor B

    Answer A is incorrect. A wet compression test is used to check for ring and cylinder wear and will not indicate anything about the head gasket condition.

    Answer B is correct. Only Technician B is correct. A substantial increase in compression during a wet compression test indicates worn rings and cylinder bores. The oil temporarily seals the rings on the cylinder, wall increasing compression.

    Answer C is incorrect. Only Technician B is correct.

    Answer D is incorrect. Technician B is correct.

10. An oil pan is being reinstalled on the engine. Technician A says that the oil pan can be sealed using RTV. Technician B says that a new gasket may be installed. Who is correct?

    TASK C.16

    A. A only
    B. B only
    C. Both A and B
    D. Neither A nor B

    Answer A is incorrect. Technician B is also correct.

    Answer B is incorrect. Technician A is also correct.

    Answer C is correct. RTV is a gasket maker, so RTV or a new gasket could be installed.

    Answer D is incorrect. Both Technicians are correct.

TASK D.10

11. Technician A says that all coolants are the same if they are ethylene glycol-based. Technician B says different coolant types can be mixed with no problems. Who is correct?

    A. A only

    B. B only

    C. Both A and B

    D. Neither A nor B

    Answer A is incorrect. All automotive coolants used by OEM manufacturers are ethylene glycol-based. The additives in each type, which may be designated by color, are different.

    Answer B is incorrect. If different types of coolant formulas are mixed, the protection will be lowered and the mixture may cause gelling in the cooling system.

    Answer C is incorrect. Neither Technician is correct.

    Answer D is correct. Neither Technician is correct. Although all OEM automotive coolants are ethylene glycol-based, the additive package is different. Manufacturers determine which additives will give the best protection and long life for their engines. The different additive packages are indicated by dye added to the coolant. Some of the different additive packages are not compatible and may cause reduced protection if mixed together. It is always best to use OEM recommended coolant.

TASK B.7

12. Which of the following tools would be used to measure valve-to-guide clearance?

    A. A micrometer and feeler gauges

    B. A dial caliper and feeler gauges

    C. A feeler gauge and machinist's rule

    D. A small hole gauge and a micrometer

    Answer A is incorrect. A feeler gauge is not suitable for this measurement.

    Answer B is incorrect. A feeler gauge is not suitable for this measurement.

    Answer C is incorrect. A feeler gauge is not suitable for this measurement and a machinist's rule is not a suitable measuring tool for this small of a measurement.

    Answer D is correct. A small hole gauge is used to transfer the guide diameter to a micrometer for the measurement.

TASK C.11

13. Technician A says before disassembly of the engine block begins, all main bearing caps must be marked for their location and orientation, if the manufacturer has not already done so. Technician B says cracked design connecting rod caps should be marked with the original orientation and cylinder. Who is correct?

    A. A only

    B. B only

    C. Both A and B

    D. Neither A nor B

    Answer A is incorrect. Technician B is also correct.

    Answer B is incorrect. Technician A is also correct.

    Answer C is correct. Both Technicians are correct. Main bearing caps are not interchangeable and cannot be reversed when reinstalled or binding of the crankshaft may occur. A cracked connecting rod cap will only fit the rod it came with.

    Answer D is incorrect. Both Technicians are correct.

TASK A.2

14. The vehicle has an intermittent no-crank condition; the starter relay has been replaced. Which of the following is the LEAST LIKELY cause?

    A. Loose connection at the battery

    B. Worn neutral safety switch

    C. Open circuit in the battery positive wire on the starter

    D. Worn ignition switch

    Answer A is incorrect. A loose connection at the battery may provide current flow intermittently.

    Answer B is incorrect. A worn neutral safety switch may cause intermittent contact.

    Answer C is correct. An open circuit in the battery positive wire on the starter will cause a no-crank condition all the time.

    Answer D is incorrect. A worn ignition switch contact may provide intermittent continuity.

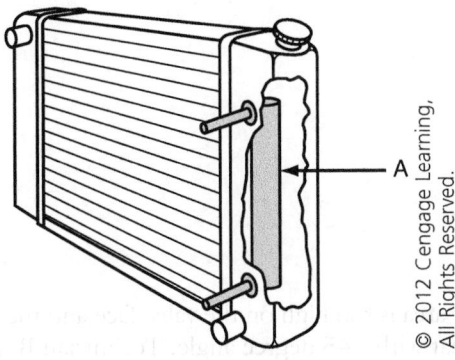

A

© 2012 Cengage Learning, All Rights Reserved.

TASK D.6

15. In the figure above, Technician A says A is an auxiliary heater. Technician B says A is an automatic transmission fluid cooler. Who is correct?

    A. A only

    B. B only

    C. Both A and B

    D. Neither A nor B

    Answer A is incorrect. A is a transmission cooler in the radiator.

    Answer B is correct. Only Technician B is correct. A is an automatic transmission cooler.

    Answer C is incorrect. Only Technician B is correct.

    Answer D is incorrect. Technician B is correct.

TASK A.3

16. Which of the following is the LEAST LIKELY location for an engine oil leak?

    A. Oil pan gasket
    B. Valve cover gasket
    C. Oil pressure sending unit
    D. Upper intake manifold gasket

    Answer A is incorrect. An oil pan gasket is a common place for an oil leak.

    Answer B is incorrect. Valve cover gaskets are common locations for oil leaks.

    Answer C is incorrect. A defective oil pressure sending unit is a common leak point.

    Answer D is correct. The upper intake manifold gasket does not carry oil through it; this would not be a location for an oil leak.

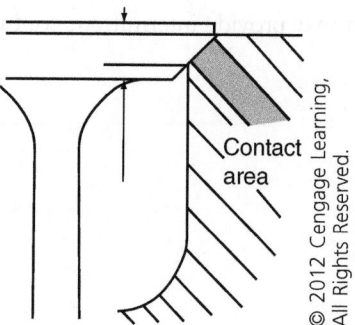

Contact area

© 2012 Cengage Learning, All Rights Reserved.

TASK B.8

17. In the figure above, the seat-to-valve contact area is too high on the valve face and the seat is too wide. Technician A says to narrow the seat with a 45 degree angle. Technician B says to top cut the seat with a 30 degree angle. Who is correct?

    A. A only
    B. B only
    C. Both A and B
    D. Neither A nor B

    Answer A is incorrect. Cutting the seat with a 45 degree angle would widen the seat more.

    Answer B is correct. Only Technician B is correct. Top cutting the seat with a 30 degree angle would move the contact down the face of the valve and narrow the seat at the same time.

    Answer C is incorrect. Only Technician B is correct.

    Answer D is incorrect. Technician B is correct.

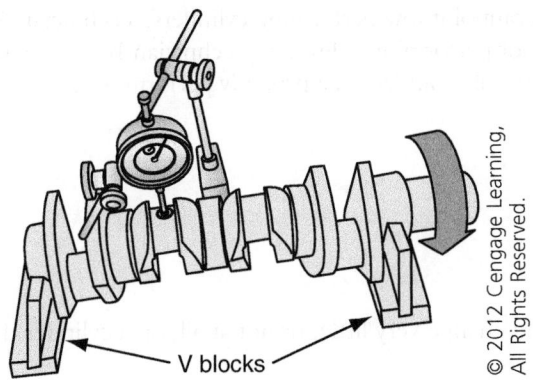

V blocks

© 2012 Cengage Learning, All Rights Reserved.

18. The figure above shows a technician:

    A. Checking journal diameter.

    B. Polishing the journal.

    C. Checking a crankshaft for runout.

    D. Demagnetizing the crankshaft.

TASK C.5

Answer A is incorrect. The journal diameter is checked with a micrometer.

Answer B is incorrect. The journal would be polished with a belt sander.

Answer C is correct. The technician is checking the runout of the crankshaft looking for warpage.

Answer D is incorrect. The crankshaft should not need demagnetizing.

19. Technician A says a V-belt that is too tight can wear out water pump bearings. Technician B says that a V-belt that is too tight can cause the #1 crank journal bearing to wear out on the upper half. Who is correct?

    A. A only

    B. B only

    C. Both A and B

    D. Neither A nor B

TASK D.7

Answer A is incorrect. Technician B is also correct.

Answer B is incorrect. Technician A is also correct.

Answer C is correct. Both Technicians are correct. A V-belt that is overtightened pulls up on the crankshaft and can cause the upper main bearing to wear out. It will also put too much pressure on one side of the water pump bearing, causing early failure.

Answer D is incorrect. Both Technicians are correct.

TASK A.7

20. A power balance test is being done to pinpoint low performing cylinders. Technician A says there will be a drop in RPM on the poor performing cylinders. Technician B says the results should be within +/− 10 percent across all cylinders of a properly performing engine. Who is correct?

   A. A only
   B. B only
   C. Both A and B
   D. Neither A nor B

   Answer A is incorrect. The RPM would change very little, or not at all, on a cylinder that is not performing properly.

   Answer B is correct. Only Technician B is correct. The expected results for all cylinders should be within +/− 10 percent of each other.

   Answer C is incorrect. Only Technician B is correct.

   Answer D is incorrect. Technician B is correct.

TASK C.11

21. Technician A says all piston pins are press fit into the connecting rod. Technician B says some piston pins are full floating. Who is correct?

   A. A only
   B. B only
   C. Both A and B
   D. Neither A nor B

   Answer A is incorrect. Only some piston pins are press fit, not all.

   Answer B is correct. Only Technician B is correct. Some piston pins are full-floating pins. They are held in place by lock rings in the piston.

   Answer C is incorrect. Only Technician B is correct.

   Answer D is incorrect. Technician B is correct.

TASK A.9

22. A cylinder leakage test has been performed and one cylinder failed with a leakage of over 80 percent. Technician A says the leakage could be into the cooling system. Technician B says testing for combustion gases in the radiator would confirm the diagnosis. Who is correct?

   A. A only
   B. B only
   C. Both A and B
   D. Neither A nor B

   Answer A is incorrect. Technician B is also correct.

   Answer B is incorrect. Technician A is also correct.

   Answer C is correct. Both Technicians are correct. A loss of pressure during a cylinder leakage test could be caused by leakage into the cooling system through a blown head gasket. Testing the radiator for combustion gases while the engine is running would confirm the diagnosis.

   Answer D is incorrect. Both Technicians are correct.

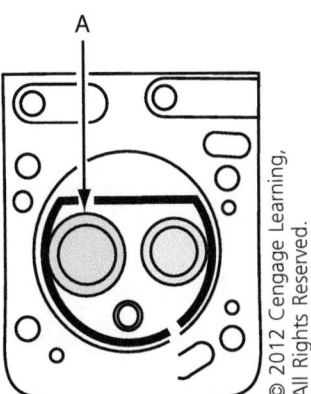

© 2012 Cengage Learning, All Rights Reserved.

23. Technician A says that A in the figure above is an exhaust valve. Technician B says the exhaust valve is always smaller than the intake valve. Who is correct?

    A. A only
    B. B only
    C. Both A and B
    D. Neither A nor B

TASK B.8

Answer A is incorrect. A in the figure is the intake valve.

Answer B is correct. Only Technician B is correct. The exhaust valve opens under pressure to empty the cylinder so it does not have to be as large as the intake.

Answer C is incorrect. Only Technician B is correct.

Answer D is incorrect. Technician B is correct.

24. Technician A says that most cylinder wear occurs in the center of piston ring travel. Technician B says that most cylinder wear occurs at the bottom of piston ring travel. Who is correct?

    A. A only
    B. B only
    C. Both A and B
    D. Neither A nor B

TASK C.4

Answer A is incorrect. Most cylinder wear occurs at the top of ring travel.

Answer B is incorrect. Most cylinder wear occurs at the top of ring travel.

Answer C is incorrect. Neither Technician is correct.

Answer D is correct. Neither Technician is correct. The center of ring travel is where the least cylinder wear occurs and the bottom of ring travel will show some wear, but the top of ring travel is where the most wear occurs. At the top of the cylinder, the piston is changing direction under combustion pressures, which will cause more pressure and wear at the top of ring travel.

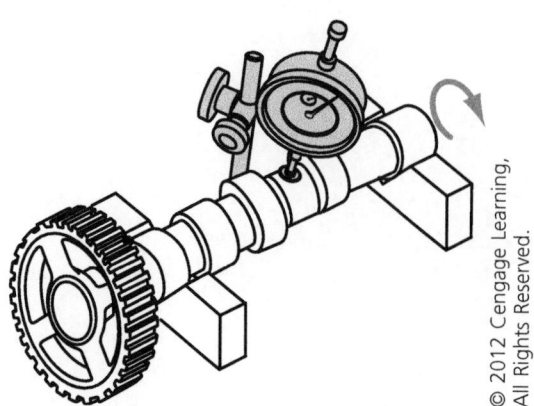

© 2012 Cengage Learning,
All Rights Reserved.

TASK B.13

25. In the figure above, Technician A says the camshaft lobe lift is being checked. Technician B says the camshaft is being checked for runout. Who is correct?

A. A only

B. B only

C. Both A and B

D. Neither A nor B

Answer A is incorrect. The technician is checking camshaft runout; the lobes are next to the journal he is checking.

Answer B is correct. Only Technician B is correct. The Technician is checking camshaft runout.

Answer C is incorrect. Only Technician B is correct.

Answer D is incorrect. Technician B is correct.

TASK A.6

26. With a vacuum gauge hooked to the engine, rapid fluctuation of the needle, from 14–18 in. Hg, is observed; this increases with RPM. Technician A says the intake manifold is restricted. Technician B says the valve springs may be weak or broken. Who is correct?

A. A only

B. B only

C. Both A and B

D. Neither A nor B

Answer A is incorrect. Intake manifold restrictions, before the port used, may cause a higher than normal and steady gauge reading.

Answer B is correct. Only Technician B is correct. Weak or broken valve springs will not close the valves properly and vacuum fluctuates rapidly.

Answer C is incorrect. Only Technician B is correct.

Answer D is incorrect. Technician B is correct.

27. An engine overheats in stop-and-go traffic but does not overheat on the highway. Technician A says a defective radiator cap may be the cause. Technician B says an inoperative electric cooling fan may be the cause. Who is correct?

TASK D.6

A. A only

B. B only

C. Both A and B

D. Neither A nor B

Answer A is incorrect. Technician B is also correct.

Answer B is incorrect. Technician A is also correct.

Answer C is correct. Both Technicians are correct. A radiator cap that will not hold its rated pressure will allow the coolant to boil below normal operating temperature; the resulting loss of coolant will cause overheating. A cooling fan is needed for airflow in stop-and-go traffic, but on the highway, ram air will carry the heat away.

Answer D is incorrect. Both Technicians are correct.

28. The LEAST LIKELY cause of excessive blue smoke from the exhaust of a turbocharged engine is:

TASK E.3

A. Worn piston rings.

B. Bad valve stem seals.

C. A PCV valve stuck in the open position.

D. Worn turbocharger seals.

Answer A is incorrect. Worn piston rings will allow oil to enter the combustion chamber causing blue smoke in the exhaust.

Answer B is incorrect. Bad valve stem seals will allow oil in the combustion chamber.

Answer C is correct. A stuck open PCV valve will not cause oil in the combustion chamber.

Answer D is incorrect. Worn turbocharger seals on the intake side will allow oil to be drawn into the combustion chamber.

29. Technician A says that a worn valve guide should be reconditioned or replaced before the valve seats are reconditioned. Technician B says the valve seats should be replaced before repairing the valve guide. Who is correct?

TASK B.7

A. A only

B. B only

C. Both A and B

D. Neither A nor B

Answer A is correct. Only Technician A is correct. Valve guides should be repaired first. The valve guide serves as the center of the seat during reconditioning.

Answer B is incorrect. The valve guide should be repaired first as it is used as the center point for seat replacement or reconditioning.

Answer C is incorrect. Only Technician A is correct.

Answer D is incorrect. Technician A is correct.

TASK E.5

30. All of the following are symptoms of a stuck open PCV valve EXCEPT:

   A. Rough engine idle.
   B. A lean air/fuel ratio.
   C. Blowby gases in the air filter.
   D. The engine stalling.

   Answer A is incorrect. A stuck open PCV valve allows unmetered air delivery to the intake.

   Answer B is incorrect. A stuck open PCV valve allows unmetered air into the intake, which will lean out the air/fuel mixture in one or more cylinders.

   Answer C is correct. A stuck open PCV valve will not cause blowby gases in the air filter: a stuck closed valve may.

   Answer D is incorrect. A stuck open PCV valve allows unmetered air into the intake, which may lean out the air/fuel mixture to the point it is unburnable.

TASK A.9

31. An engine miss is being diagnosed using a cylinder leakage test. Technician A says that a 20 percent leakage is acceptable. Technician B says air coming out the intake manifold indicates a cracked cylinder head. Who is correct?

   A. A only
   B. B only
   C. Both A and B
   D. Neither A nor B

   Answer A is correct. Only Technician A is correct. The acceptable leakage from a cylinder is 20 percent. This allows for normal leakage past ring end gaps.

   Answer B is incorrect. Air coming out the intake manifold indicates leaking intake valves.

   Answer C is incorrect. Only Technician A is correct.

   Answer D is incorrect. Technician A is correct.

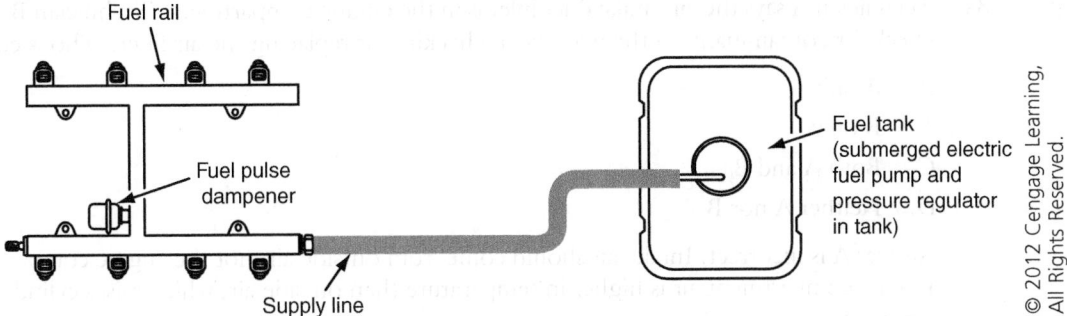

© 2012 Cengage Learning, All Rights Reserved.

32. Technician A says the figure above shows a returnless fuel system. Technician B says the fuel pump and pressure regulator are in the gas tank. Who is correct?

    A.   A only
    B.   B only
    C.   Both A and B
    D.   Neither A nor B

    TASK E.1

    Answer A is incorrect. Technician B is also correct.

    Answer B is incorrect. Technician A is also correct.

    Answer C is correct. Both Technicians are correct. The system shown is a returnless system, which has the pump and pressure regulator in the gas tank.

    Answer D is incorrect. Both Technicians are correct.

33. Technician A says typical piston-to-bore clearance is 0.020 to 0.030 inch. Technician B says typical piston-to-bore clearance is 0.001 to 0.002 inch. Who is correct?

    A.   A only
    B.   B only
    C.   Both A and B
    D.   Neither A nor B

    TASK C.12

    Answer A is incorrect. Typical piston-to-bore clearance is 0.001 to 0.002 inch.

    Answer B is correct. Only Technician B is correct. Typical piston-to-bore clearance is 0.001 to 0.002 inch; this is calculated by measuring the bore and the piston and subtracting.

    Answer C is incorrect. Only Technician B is correct.

    Answer D is incorrect. Technician B is correct.

TASK E.2

34. Technician A says the air intake duct inlet is in the engine compartment. Technician B says to check for contaminants in the inlet when checking or replacing the air filter. Who is correct?

  A. A only

  B. B only

  C. Both A and B

  D. Neither A nor B

Answer A is incorrect. Intake air should come from outside air, not the engine compartment. Engine compartment air is higher in temperature than outside air, which raises cylinder temperature.

Answer B is correct. Only Technician B is correct. If the inlet is not cleaned when the filter is changed, the contaminants in it may block the new air filter.

Answer C is incorrect. Only Technician B is correct.

Answer D is incorrect. Technician B is correct.

TASK B.16

35. Technician A says stuck or sticky valves may cause bent valves. Technician B says a timing belt off by three teeth may cause bent valves. Who is correct?

  A. A only

  B. B only

  C. Both A and B

  D. Neither A nor B

Answer A is incorrect. Technician B is also correct.

Answer B is incorrect. Technician A is also correct.

Answer C is correct. Both Technicians are correct. Any time a valve fails to close at the proper time, the piston may contact it. A sticky or stuck valve is slow to close. If the camshaft and crankshaft are out of synchronization, the piston may contact a valve.

Answer D is incorrect. Both Technicians are correct.

TASK D.3

36. Technician A says an engine oil cooler may be located in one of the radiator tanks. Technician B says a transmission oil cooler may be located in one of the radiator tanks. Who is correct?

  A. A only

  B. B only

  C. Both A and B

  D. Neither A nor B

Answer A is incorrect. Technician B is also correct.

Answer B is incorrect. Technician A is also correct.

Answer C is correct. Both Technicians are correct. An oil cooler tank and/or a transmission cooler may be mounted inside the radiator side tanks.

Answer D is incorrect. Both Technicians are correct.

37. Technician A says the plug wires on a failed ignition coil pack must be checked for defects. Technician B says most coil packs are considered waste spark systems. Who is correct?

    A. A only
    B. B only
    C. Both A and B
    D. Neither A nor B

    TASK E.6

    Answer A is incorrect. Technician B is also correct.

    Answer B is incorrect. Technician A is also correct.

    Answer C is correct. Both Technicians are correct. An open plug wire may cause the coil pack to internally arc to the primary winding, causing failure of the coil and possibly the module. Most coil packs are waste spark systems in that they fire two spark plugs at one time, one for ignition and one for waste spark on the exhaust stroke.

    Answer D is incorect. Both Technicians are correct.

38. Technician A says that when adjusting valves on an engine with mechanical adjustment, the piston must be at BDC on the exhaust stroke. Technician B says some mechanical valve adjustment is done by changing removable lash pads on a bucket. Who is correct?

    A. A only
    B. B only
    C. Both A and B
    D. Neither A nor B

    TASK B.12

    Answer A is incorrect. When mechanically adjusting valves, it must be done with the piston at TDC compression stroke to ensure all valves for that cylinder are closed.

    Answer B is correct. Only Technician B is correct. Some engines' camshafts may act on a lash pad and bucket assembly, which directly contacts the valve tip.

    Answer C is incorrect. Only Technician B is correct.

    Answer D is incorrect. Technician B is correct.

39. The engine cranks but will not start. Which of the following is not a requirement for the engine to start and run?

    A. Compression
    B. Oil pressure
    C. Fuel pressure
    D. Spark

    TASK A.2

    Answer A is incorrect. Compression is required for an engine to start.

    Answer B is correct. Oil pressure is developed when the engine starts running.

    Answer C is incorrect. Fuel pressure and delivery is required for an engine to start.

    Answer D is incorrect. Spark is required to ignite the fuel.

TASK A.1

40. A technician begins his diagnostic procedure to repair a vehicle. Which of the following would be the LEAST LIKELY steps he would take?

A. Road test the vehicle.

B. Question the customer for more information regarding the problem.

C. Make sure the original complaint is fixed.

D. Start with the most difficult tests.

Answer A is incorrect. A technician should test drive the vehicle to verify the complaint.

Answer B is incorrect. It is good practice to question the customer to get all the information regarding the problem that he can provide.

Answer C is incorrect. The original complaint must be fixed to ensure customer satisfaction.

Answer D is correct. During diagnosis, the technician will start with the simple and work his way through to the difficult tests. Most of the time the problem will be diagnosed and corrected before he gets to the difficult tests.

TASK C.6

41. Technician A says the main bearing bores are checked for alignment with a straightedge and a feeler gauge. Technician B says main bearing bores that are out of alignment can be corrected by line boring. Who is correct?

A. A only

B. B only

C. Both A and B

D. Neither A nor B

Answer A is incorrect. Technician B is also correct.

Answer B is incorrect. Technician A is also correct.

Answer C is correct. Both Technicians are correct. The proper tools to check main bearing bore alignment are the straightedge and feeler gauge; line boring would straighten out of alignment bores.

Answer D is incorrect. Both Technicians are correct.

TASK C.2

42. Technician A says the upper oil galleys in the block may be sealed with pipe plugs. Technician B says the oil galley plugs must be removed to properly clean the galleys. Who is correct?

A. A only

B. B only

C. Both A and B

D. Neither A nor B

Answer A is incorrect. Technician B is also correct.

Answer B is incorrect. Technician A is also correct.

Answer C is correct. Many upper oil galleys are sealed with pipe plugs that must be removed during cleaning to make sure the galleys are completely clean.

Answer D is incorrect. Both Technicians are correct.

43. Technician A says that valve train related noise could be caused by a low oil level. Technician B says that valve train related noise could be caused by valves that need to be adjusted. Who is correct?

TASK A.4

A. A only

B. B only

C. Both A and B

D. Neither A nor B

Answer A is incorrect. Technician B is also correct.

Answer B is incorrect. Technician A is also correct.

Answer C is correct. Both Technicians are correct. A low oil level will reduce the amount of oil pumped to the top of the engine, leaving the valves with no lubrication. If a valve train has excessive clearance between components, it will cause noise until corrected.

Answer D is incorrect. Both Technicians are correct.

44. A technician is changing oil in a customer's vehicle and needs to know what weight oil to put in. Technician A says the weight of oil does not matter as long as it is multigrade oil. Technician B says the technician should look on the oil filler cap or in the owner's manual. Who is correct?

TASK D.4

A. A only

B. B only

C. Both A and B

D. Neither A nor B

Answer A is incorrect. Refilling the crankcase with the wrong weight oil may cause higher than specified oil pressure and void a customer's warranty on the engine.

Answer B is correct. Only Technician B is correct. Most manufacturers emboss the proper oil weight on the oil fill cap and specify oil weight in the owner's manual.

Answer C is incorrect. Only Technician B is correct.

Answer D is incorrect. Technician B is correct.

45. Technician A says that when removing a cylinder head from an OHC engine, the camshaft may have to be removed first. Technician B says the cylinder head should be cold before removal. Who is correct?

TASK B.1

A. A only

B. B only

C. Both A and B

D. Neither A nor B

Answer A is incorrect. Technician B is also correct.

Answer B is incorrect. Technician A is also correct.

Answer C is correct. Both Technicians are correct. Many OHC engines require removal of the camshaft to access the head bolts. The cylinder head should be cold before removal, particularly with an aluminum head, to prevent inducing warpage when the head bolts are removed.

Answer D is incorrect. Both Technicians are correct.

TASK A.5

46. The customer is concerned about his oil pressure gauge dropping after the vehicle is driven for about 20 miles, although when he starts up, it is normal. When the dipstick is checked after shutdown, it shows below the add mark. When the dipstick is checked after sitting, the oil level shows full. Which of the following is the most likely cause?

A.   Worn camshaft bearings

B.   Leaking rear main seal

C.   Blocked oil return holes in the head

D.   Weak oil pump

Answer A is incorrect. Worn camshaft bearings may affect oil pressure, but not oil level.

Answer B is incorrect. If the rear main seal was leaking, the oil level would stay low from oil loss.

Answer C is correct. Blocked return holes in the head will hold the oil in the valve cover and slowly allow it to drain back into the pan.

Answer D is incorrect. A weak oil pump will cause low oil pressure, not oil use.

TASK C.7

47. Technician A says that during engine reassembly, all rod bearing clearances should be checked. Technician B says the rod bearing inserts should extend slightly above the rod and cap; this is called bearing crush. Who is correct?

A.   A only

B.   B only

C.   Both A and B

D.   Neither A nor B

Answer A is incorrect. Technician B is also correct.

Answer B is incorrect. Technician A is also correct.

Answer C is correct. Both Technicians are correct. All rod bearing clearances should be checked to ensure proper bearing-to-journal clearance. Bearing crush ensures that the bearing fits tight in the rod when assembled and will not rock.

Answer D is incorrect. Both Technicians are correct.

TASK A.8

48. While performing a cranking compression test on a 4-cylinder engine, the technician notes that one cylinder has a pressure reading of 60 psi, 44 kPa, while the others have a reading of 135 psi, 931 kPa. Technician A says performing a cylinder leakage test will indicate where the pressure is leaking. Technician B says the vehicle has worn valve guides. Who is correct?

A.   A only

B.   B only

C.   Both A and B

D.   Neither A nor B

Answer A is correct. Only Technician A is correct. The next step for narrowing down the cause for low compression on one cylinder would be the cylinder leakage test.

Answer B is incorrect. Worn valve guides will not cause a loss of compression.

Answer C is incorrect. Only Technician A is correct.

Answer D is incorrect. Technician A is correct.

49. Which of the following could cause symptoms of overheating even though the engine temperature is normal?

TASK D.13

   A. A stuck closed thermostat
   B. A defective radiator cap
   C. A defective temperature sending unit
   D. A missing thermostat

   Answer A is incorrect. A stuck closed thermostat will make the engine actually overheat.

   Answer B is incorrect. A defective radiator cap may cause an actual overheat condition.

   Answer C is correct. If the temperature sending unit has gone bad, it can cause the gauge to register an overheat condition when, in fact, the engine is at the normal operating temperature.

   Answer D is incorrect. A missing thermostat will cause engine overcooling.

50. The customer complains of a loud thump when he accelerates from a stop. Technician A says it could be caused by worn crankshaft thrust bearings. Technician B says it could be caused by a cracked flexplate. Who is correct?

TASK A.4

   A. A only
   B. B only
   C. Both A and B
   D. Neither A nor B

   Answer A is correct. Only Technician A is correct. Excessively worn thrust bearings will allow the crankshaft to shift forward during acceleration, causing a thump when the crankshaft comes up against the main bearing thrust surface.

   Answer B is incorrect. A cracked flexplate will typically sound like a sharp pop or knock as it flexes at the crack.

   Answer C is incorrect. Only Technician A is correct.

   Answer D is incorrect. Technician A is correct.

PREPARATION EXAM ANSWER SHEET FORMS

ANSWER SHEET

1. _____	21. _____	41. _____
2. _____	22. _____	42. _____
3. _____	23. _____	43. _____
4. _____	24. _____	44. _____
5. _____	25. _____	45. _____
6. _____	26. _____	46. _____
7. _____	27. _____	47. _____
8. _____	28. _____	48. _____
9. _____	29. _____	49. _____
10. _____	30. _____	50. _____
11. _____	31. _____	
12. _____	32. _____	
13. _____	33. _____	
14. _____	34. _____	
15. _____	35. _____	
16. _____	36. _____	
17. _____	37. _____	
18. _____	38. _____	
19. _____	39. _____	
20. _____	40. _____	

ANSWER SHEET

1. _____	21. _____	41. _____
2. _____	22. _____	42. _____
3. _____	23. _____	43. _____
4. _____	24. _____	44. _____
5. _____	25. _____	45. _____
6. _____	26. _____	46. _____
7. _____	27. _____	47. _____
8. _____	28. _____	48. _____
9. _____	29. _____	49. _____
10. _____	30. _____	50. _____
11. _____	31. _____	
12. _____	32. _____	
13. _____	33. _____	
14. _____	34. _____	
15. _____	35. _____	
16. _____	36. _____	
17. _____	37. _____	
18. _____	38. _____	
19. _____	39. _____	
20. _____	40. _____	

ANSWER SHEET

1. _____	21. _____	41. _____
2. _____	22. _____	42. _____
3. _____	23. _____	43. _____
4. _____	24. _____	44. _____
5. _____	25. _____	45. _____
6. _____	26. _____	46. _____
7. _____	27. _____	47. _____
8. _____	28. _____	48. _____
9. _____	29. _____	49. _____
10. _____	30. _____	50. _____
11. _____	31. _____	
12. _____	32. _____	
13. _____	33. _____	
14. _____	34. _____	
15. _____	35. _____	
16. _____	36. _____	
17. _____	37. _____	
18. _____	38. _____	
19. _____	39. _____	
20. _____	40. _____	

ANSWER SHEET

1. _____	21. _____	41. _____
2. _____	22. _____	42. _____
3. _____	23. _____	43. _____
4. _____	24. _____	44. _____
5. _____	25. _____	45. _____
6. _____	26. _____	46. _____
7. _____	27. _____	47. _____
8. _____	28. _____	48. _____
9. _____	29. _____	49. _____
10. _____	30. _____	50. _____
11. _____	31. _____	
12. _____	32. _____	
13. _____	33. _____	
14. _____	34. _____	
15. _____	35. _____	
16. _____	36. _____	
17. _____	37. _____	
18. _____	38. _____	
19. _____	39. _____	
20. _____	40. _____	

ANSWER SHEET

1. _____	21. _____	41. _____
2. _____	22. _____	42. _____
3. _____	23. _____	43. _____
4. _____	24. _____	44. _____
5. _____	25. _____	45. _____
6. _____	26. _____	46. _____
7. _____	27. _____	47. _____
8. _____	28. _____	48. _____
9. _____	29. _____	49. _____
10. _____	30. _____	50. _____
11. _____	31. _____	
12. _____	32. _____	
13. _____	33. _____	
14. _____	34. _____	
15. _____	35. _____	
16. _____	36. _____	
17. _____	37. _____	
18. _____	38. _____	
19. _____	39. _____	
20. _____	40. _____	

Glossary

Air conditioning The process of adjusting or regulating, by heating or cooling, the quality, temperature, and humidity of air.

Aluminum A nonferrous metal that is light in weight yet can be stronger than steel when mixed with the proper alloys. It is easily cast and machined.

Balance shaft A shaft with counterweights designed to prevent vibration of rotating parts.

Battery A device for storing electrical energy in chemical form.

Bearing crush Occurs when the bearing cap is tightened, the slightly protruding bearing shell ends touch, and are forced together. This provides a tight fit in the bore for the bearings and prevents bearing rock.

Belt A device used to drive the water pump and other accessory power-driven devices.

Block deck The flat surface of the main casting of an engine on which the head attaches.

Cam An abbreviation for camshaft: a device having lobes, driven by the crankshaft via gears, a chain, or a belt that opens and closes the intake and exhaust valves.

Cam follower A term used for valve lifter: a hydraulic or mechanical device, in the valve train, that rides on the camshaft lobe to lift the valve off its seat.

Camshaft journal That part of the camshaft that turns in a bearing.

Cast iron A term used for various cast ferrous alloys containing at least 2 percent carbon. It is used for many different parts on vehicles.

Catalytic converter An automotive exhaust system component, made of stainless steel, containing a catalyst that reduces hydrocarbons, carbon monoxide, and nitrogen oxides present in the engine exhaust gases.

Coil A term used to describe a spring or an electrical device using many turns (coils) of wire such as springs, ignition coils, solenoids, and relays.

Connecting rod bearing The bearing of a connecting rod that rotates on the crankshaft.

Cooling system The system that circulates coolant through the engine to dissipate its heat.

Counterbalance shafts One or more rotating shafts found on some engines to counteract the natural vibrations of other rotating parts, such as the crankshaft, in that engine. This balance, or counterbalance, shaft usually turns at twice the crankshaft speed and must be timed properly.

Crankshaft The revolving part of a unit that has the function of delivering power or work from the reciprocating motion. Engines have crankshafts, as do air conditioning compressors and air compressors.

Crankshaft sensor An electronic device used to send crankshaft rotating information to the computer.

Cross–firing A condition whereby spark plugs fire out of turn, usually caused by poor spark plug wire insulation.

Cylinder head That part of the engine that covers the cylinders and pistons.

Cylinder out of round The difference in the cylinder inside diameter when measured in the same plane, 90 degrees apart.

Cylinder taper The difference in cylinder inside diameter from top to bottom.

Cylinder wall A term used for cylinder bore, the inside diameter of a cylinder.

Distributor A device used on many engines to direct high-voltage electrical energy from the coil to the spark plugs.

Elongated Not round, but egg-shaped.

End-play A term used to describe a spacing or clearance involved with a moving part. Crankshafts and camshafts require end-play measurements and must be set to manufacturer's specifications during engine assembly.

Engine oil A lubricant formulated for use in an engine.

Engine oil cooler A device used on some high performance engines, police packages, taxis, trucks, turbo-equipped engines, and diesel engines to prevent the engine oil from overheating. These work by using a heat exchanger exposed to air flow or engine coolant.

Exhaust conditioning The burned and unburned gases that remain after combustion.

Exhaust pipe A pipe that connects the exhaust manifold to the muffler or catalytic converter. It is made of heavier material than tailpipes, and is sometimes double-layered.

Face-to-seat As in valve face-to-valve seat contact, this refers to the actual sealing area of the valve and seat. It must be the size and angle specified by the engine manufacturer to properly seal the combustion chamber and have a long service life.

Firing order The order in which the engine cylinders fire and deliver power.

Flywheel A round, heavy metal plate attached to the crankshaft of an engine that helps smooth out power strokes and gives the rotating crankshaft momentum to smoothly get to the next power stroke. The clutch and pressure plate help the flywheel transmit power to the drive wheels.

Frozen A mechanical problem developed from a lack of oil or broken internal parts that prevents motion, such as an engine rotating.

Garter spring A small spring placed behind the lip of a lip seal to maintain contact with the rotating part; most often associated with oil seals.

Gap A space between two adjacent parts or surfaces.

Head That part of an engine that covers the top of the cylinders and pistons.

Heat shield Devices used in many places on today's vehicles. One such place is between the starter solenoid and the heat of the engine and exhaust manifold. Another is between electrical wiring or spark plug wiring and any high heat source. Heat shields are also used between the catalytic converters and the passenger compartment of the vehicle.

Hone To use abrasive materials to remove material from a surface or just to change the surface smoothness so the parts involved will work better.

Idler pulley A pulley that is used to adjust the belts on a belt-driven system.

Input A term used for the signals sent to different electronic modules about operating conditions of systems involved.

Intake ductwork All of the connecting ductwork from the throttle body out involved with getting air into the engine.

Jumped As in jumped timing chain or timing belt, resulting in a condition where the valve timing is no longer correct, and the engine will run poorly or might not start.

Keepers Key-like tapered metal locking devices used to hold valve retainers in place.

Lash The clearance between two parts.

Lash adjuster A device much like hydraulic lifters that is usually found on overhead cam engines. Its job is to maintain zero lash between the cam follower and the tip of the valve.

Line boring A machining process that ensures multiple holes that are bored in a cylinder head or block are in line or true. This allows the camshaft or crankshaft installed in these locations to turn freely and function properly. As in line bore alignment of camshaft bearing or main bearing bores, a special boring bar is used that removes metal from all cam or main bearing bores at the same time and in a straight and true line.

Line hone A boring tool used to re-establish the correct main bearing bore center line in a cylinder block.

Main bearing cap The structural device that holds the crankshaft in place in an engine block.

Muffler A device in the exhaust system used to reduce noise.

No-crank A condition like a frozen engine, a defective starter, or an electrical problem that prevents the engine from rotating when normal attempts are made to start the engine.

No-start A condition where the engine turns over normally but does not start. This could be caused by mechanical problems, fuel system problems, electrical problems, or electronic engine control problems.

Oil cooler A heat exchanger used to cool transmission or engine oil. Also may be used to cool power steering or other fluids.

Oil filter A device used to remove impurities, such as abrasive particles, from oil.

Oil pan A removable part of the engine assembly that contains the oil supply.

Out of square As in a valve spring that is not within specification when checked vertically on a flat surface and measured at a 90-degree angle. If installed in an engine, such a spring would cause side pressures on the valve and damage the valve and valve guide.

Overhead cam A camshaft that is mounted in the cylinder head.

Oxygen sensor An electronic device found in the exhaust system that measures the amount of oxygen in the exhaust stream.

Piston pin A precision ground pin, usually hollow, used to attach the connecting rod to the piston. The piston pin can be held in place by a press fit, snap rings, or bolts.

Pressure cap A cap placed on the radiator to allow regulated, above atmospheric, pressure in the cooling system.

Primed Ready or prepared.

Pulley A wheel-shaped device used in a belt-drive system to drive accessory equipment.

Ram air Air forced through the radiator, condenser, and across the engine by the forward movement of the vehicle.

Reluctor A gear-like part of an electronic ignition system. It could have the same number of teeth (or gaps) as cylinders of the engine, or it could have half the number of teeth as the engine has cylinders. As a tooth (or gap) passes by a pickup coil or crankshaft sensor, the magnetic field changes, and a trigger signal is sent to an electronic control module.

Reluctor ring A gear-like part of the electronic ignition system.

Rotators As in valve-rotator devices that cause the valves in the cylinder head to rotate slightly each time the valve closes. This helps keep the valve seat and valve face clean. Because a different part of the valve face contacts the valve seat each time it closes, the valve also operates at a cooler temperature. This greatly extends valve and seat life.

Rotor A part of the ignition distributor that rotates inside the cap and transfers ignition coil secondary electrical energy from the center tower to the individual spark plug wires.

RTV An abbreviation for Room Temperature Vulcanizing; the trade name for a rubber-like sealing compound.

Select fit As in the main bearings of an engine. Many of today's engines use select fit main bearings. This means the main bearings are no longer all one standard size, or all 0.010 inch

(0.254 mm) undersized. The manufacturer mixes and matches bearing halves of small increments of as little as 0.003 inch (0.076 mm) to ensure better oil and noise control at the crankshaft area.

Selective thrust washer A washer or spacer that is furnished in different sizes to facilitate end-play and preload settings.

Skirt A term used to describe the lower part of an engine piston. The skirt contacts the cylinder wall, helps the piston travel in a straight line, and prevents piston slap.

Solenoid An electromechanical device used for a push-pull operation.

Span A term used for the length or space between two parts; as in the span between the air conditioner compressor belt pulley and the alternator pulley is too great. This condition can cause the belts to fly off at high engine speeds or during rapid acceleration.

Spark plug A component of the ignition system that delivers the high-voltage spark to the combustion chamber.

Starter drive The part of the starter motor that engages the ring gear or the flywheel, flexplate, or torque converter, and rotates the engine on start up.

Starter motor The small electric motor that is used to crank (start) an engine.

Starter solenoid The part that causes the starter drive to engage the flywheel when starting an engine.

Tailpipe The pipe from the muffler or catalytic converter that carries the exhaust gases away from the passenger compartment.

Temperature sensor A term used for various temperature sensing switches or variable resistors on today's vehicles. They can be used to turn cooling fans on and off, to operate dash gauges, to control air conditioners, and to furnish inputs to engine and transmission control computers.

Tensioner A device used with timing belts and timing chains that maintains a constant pressure on the belt or chain to minimize wear and noise, and to take up slack or lost motion. They may be fixed and require periodic adjustment, or they may be automatically adjusted by spring tension or engine oil pressure. Tensioners are also used on accessory drive belts.

Timing belt The belt through which the crankshaft drives the camshaft(s) in an overhead cam engine.

Torque wrench A specially designed tightening device that indicates the amount of torque being exerted on a fastener to enable threaded parts to be tightened to a specified amount.

Torque-to-yield (TTY) A term used to describe a common method of tightening fasteners on many of today's engines. The procedure is to tighten the fasteners to a fairly low pounds-feet value, then to turn each fastener (in proper tightening sequence) a specified number of degrees. This process is usually used on head bolts, main bearing bolts, and rod-bearing bolts. In most cases, new fasteners are required for every reassembly.

True A term used in the automotive industry to signify a part or system is correct and is within specifications. As in the cylinder head gasket sealing surface is true; i.e., it is not warped, scratched, broken, or otherwise damaged.

Valve float A condition that occurs when the valve spring is not capable of closing the valve quickly enough. This usually happens at higher engine speeds, and is aggravated by the valve spring losing some of its tension. Improper sealing of the combustion chamber will occur and the engine will run poorly.

Valve lifter A hydraulic or mechanical device in the valve train that rides on the camshaft lobe to lift the valve off of its seat.

Valve overlap The amount of time measured in degrees during which the intake and exhaust valves are both open.

Valve rotator A device that rotates the valve while the engine is running.

Valve spring installed height The distance from the underside of the valve spring retainer to the cylinder head surface.

Valve spring retainer A device on the valve stem that holds the spring in place.

Valve train The parts making up the valve assembly and its operating mechanism.

Warpage As in a cylinder head gasket, the sealing area is no longer being straight and true. This condition will require machine shop work on the cylinder head or a different cylinder head that is not warped.

Water pump A mechanical device used to circulate coolant through the cooling system.

Witness marks Lines scribed on adjacent surfaces of mating parts, before disassembly, to ensure proper alignment when reassembled.

Notes

Notes

Notes

Notes

Notes